Bill Turnbull has worked for the BBC since 1986 and joined the BBC Breakfast team as a presenter in 2000. He lives in Buckinghamshire with his wife, three children, two black Labradors, five chickens and an awful lot of bees.

the BAD BEEKEEPERS
Club

*How I stumbled into the curious
world of bees – and became (perhaps)
a better person*

BILL TURNBULL

sphere

SPHERE

First published in Great Britain in 2010 by Sphere
Reprinted 2010 (twice)
This paperback edition published in 2011 by Sphere

A CIP catalogue record for this book
is available from the British Library.

ISBN 978-0-7515-4405-3

Typeset in Garamond by M Rules
Printed and bound in Great Britain by
Clays Ltd, St Ives plc

Sphere
An imprint of
Little, Brown Book Group
100 Victoria Embankment
London EC4Y 0DY

An Hachette UK Company
www.hachette.co.uk

www.littlebrown.co.uk

To Sesi

Contents

Introduction

Hello. My name is Bill, and I'm a Bad Beekeeper. A really Bad Beekeeper. I've done bad things with bees. Terrible things. Things you wouldn't understand unless you were a beekeeper yourself.

I still shudder at the thought of one or two of them. I couldn't put them down here. If I did, and you were even a half-decent beekeeper, you'd probably stop reading right here and now.

But I keep going. I can't help myself.

Why, I hear you ask. Why, oh why?

I just have to. It's like they're my path to redemption. Ever since a swarm arrived in our garden seventeen years ago, I'd wondered if I could be brave enough to take up the challenge. I still remember standing in awed silence as I watched (from behind the locked patio doors) a local beekeeper turn up with a pair of secateurs and a cardboard box, and take them away without any fuss. It was then that I knew I had to give it a go, although there is a second explanation involving a sick chicken called Tabasco and a trip to the vet, but I'm saving that for chapter two.

Bees can be very forgiving. They'll put up with a lot. They'll punish you when you do wrong and, believe me, in my time I have indeed been punished. But treated right, and left for long enough, their mood will improve. And like many a long-suffering woman – for bees are overwhelmingly female – they will give you another chance. Eventually.

And so I like to think, in the words of the French psychologist Émile Coué – or was it Winnie the Pooh? – 'Every day in every way I'm getting better and better'. It's been suggested that Émile Coué eventually committed suicide. But we'll not let that put us off.

I'm in the rare and privileged position of being a BBC man twice over – once with the finest broadcasting corporation the world has ever seen, where I make a living presenting the *Breakfast* programme on BBC One, and again as a founder member of the Bad Beekeepers Club. I set it up a couple of years ago with my beekeeping friend Peter Murray Russell, a kindred spirit who seemed to be almost as inept as I was when it came to keeping bees and making honey. I once found a road map in one of his hives. He offered a long and complicated technical justification, but I think he was just worried that they might not find their way home. By the way, reading this book confers on you honorary membership of this small but august association. Well, we've got to get the numbers up somehow.

Think of this book as a sort of induction course, to let you know what you can expect should you decide to venture into our fascinating world of veils and gauntlets and odd-looking hive tools. Let me show you how it all works, how the bees make the honey and how we then nick it off them; the pains and pitfalls you can encounter, as well as the joys you can experience when the job isn't even half well done.

If you're not convinced – and after reading this you might well wonder who would be – think of this as a gentle guide to demystify the process, to teach you a bit about the insects that are so important to us, and remind you how you got off lightly.

And if you're a beekeeper already, you can of course just read it and weep.

I just hope that if I keep trying, one day I might not be *such* a Bad Beekeeper. I might progress to being simply 'a bee-keeper'. Not a good one, but not a bad one either.

So, gloves on. Zip up your veil and join me, as I prise the lid off my far too imperfect career as an apiarist. And along the way, we might learn a thing or two about life itself.

Ready?

Chapter One

FORTITUDE

*In which we discover that learning
can be a painful process. Even on
the very first day.*

It wasn't the best of starts. In fact, as starts go, it was about as bad as it can get. But I suppose at least it was unforgettable. My first trip to an apiary – the place where bees live – became not so much a near disaster as a near death experience.

I had been warned. 'You will get stung,' was the mantra of the first beekeeping lesson I had been to at the Pinner & Ruislip Beekeepers Association (PRBKA), deep in the suburban outskirts of London. It was drilled into us repeatedly that beekeeping was not without its hazards, and getting stung was chief among them. And in a strange and possibly rather masochistic way, I have now been piqued often enough to conclude that getting stung really is where the fun is. Beekeepers can be odd like that.

It was midday on a warm, sunny Sunday, and I was more than ready for my first encounter with some live bees. I'd learnt the basic theory with three other keen apprentices from our teacher at the PRBKA, taking notes about colonies and hives and basic bee biology, and couldn't wait to get out there and have a go myself. In the back of my mind, I remembered that there was a risk of some kind associated with bees. But bad things would only happen if I did something to provoke them. And if I moved slowly and calmly and didn't drop anything, chances were I would emerge unscathed.

So it was more with eagerness than apprehension that I parked that Sunday afternoon in the shade outside the apiary; a paddock about the size of a football pitch, dotted with beehives and, of course, busy with bees. The sun was shining and the two experienced beekeepers who were to show me the ropes were already at work, with one of the hives open for inspection. I stood at the entrance to the apiary, where a white van was parked, and announced myself. 'Hello, yes,' came the reply. 'There's a suit for you in the van there. Oh, and just watch out for the swarm on the fencepost.'

'Ah, a swarm on the fencepost,' I repeated anxiously, noticing that said fencepost was a mere eight feet from where I was standing.

Mention the word swarm to most people and they automatically tend to cover up and recall tales of how they'd heard that the aunt of a friend of a neighbour's local newsagent was once smothered in one the size of a house and never seen again. They picture swarms as vast hostile clouds of wild flying insects, ready to attack anything and anyone who stumbles into their path. In truth, it's not really like that. Yes, swarms do consist of vast clouds of wild flying insects, but they're not usually hostile and can be fascinating to watch. After a while they settle together in a large brown clump to have a breather and work out what to do next, resting beneath the branch of a tree or, in this case, a fencepost.

There they were, doing their own thing, as I gingerly approached the van from the other side to reach in for the suit, which looks like a large white boiler suit with a veil attached. So far, so good. But the thing about swarms is that they are not static. In other words, while the main body of the colony may indeed be clustered on a fencepost, there can be any number of extraneous bees flying around it – latecomers arriving from their old home, others to-ing and fro-ing looking for new places to live in, or general hangers-on buzzing around just for the fun of it.

It was a couple of these outriders who took an interest in me as I started to put the suit on. Their interest was probably completely innocent, but it was slightly unnerving on my first time out. So, remembering my beekeeping basic training, I wandered into the shade at the side of the road. They're supposed to leave you alone in the shade, and head back to the sun. But this pair were defying convention. And clearly by now I was more than just vaguely interesting to these two, who began to hover around my head with some interest.

Today I always offer simple advice to anyone caught in this situation: 'Stand dead still, and stay dead calm.' But I know from experience that this is much more easily said than done. When some persistent little buzzer is whining around the back of your head, as they are prone to do, your every instinct is to do *anything* but stand still. Run, shake, swat, hide; anything to get away. But I can also tell you this definitely doesn't work. An Olympic athlete might just outdo a bee in a one-hundred-metre sprint, but since they can get up to twenty miles an hour in flight, even the fastest and fittest mammal would have their work cut out. As for swatting, forget it. A bee has five eyes for a start. And thanks to what's called 'flicker fusion potential', it can see at the equivalent of three hundred frames a second. If bees went to the cinema, the film would appear to them as a long sequence of still pictures. In other words, they could see your hand moving like it's a slow-motion action replay.

Hiding can do the trick, if there's a dark room close by that you could get to – preferably one that's mildly refrigerated. Sadly there wasn't one available at the Harefield Apiary that day, and I was about to forget the sage advice that I'm now giving you. I did try to stay calm for a while, until they landed on my head. Then, panic rising, I resorted to what came naturally and tried to brush them off, in the nicest possible way. Which is where we come to hair care.

Perhaps if I hadn't been working that morning I'd have been all right. I'd have come to the apiary with my 'coiffure au naturel'. But as I'd been in the studio, some gel had been applied to render me respectable for the viewing public first thing on a Sunday morning. As a result, my hair was ever so slightly sticky. Not a problem for day-to-day living, except when bees become involved. They must have liked the smell; perhaps they thought it was a delightful new form of food supply. You can imagine the conversation back at the hive. 'What have you been on today, Doris?'

'Oh, the usual – dandelion and honeysuckle. But I hear Ethel and her sister have been on some fabulous new stuff called Brylcreem. It's positively dreamy.'

The problem was that once they'd landed on it, they couldn't get off again in a hurry. Given time, they would have managed. But as they were being swatted rather frantically by an enormous paddle – my hand – they couldn't exactly hang about. So they buzzed furiously around, or rather on, my head while I brushed equally furiously with the back of my hand. Sooner or later, something had to give.

I've never actually been hit on the head by a hammer, but that day I got a pretty good idea of what it's like. (Although when I was thirteen I did suffer a sound thump to the forehead from a croquet mallet, inflicting a bloody wound which required eight painful stitches above my left eye. The joys of prep school. Fortunately, most stings are mildly less life-threatening.) When a bee stings you on the head, there's no flesh to cushion the barb, so it does feel rather like a sharp blow with a blunt object, if that's possible. Think of a hammer with a drawing pin attached. And since both bees felt threatened enough to risk their lives in their own defence, think of two hammers, both with drawing pins attached.

At least it stopped the buzzing. But there was more to come.

Eyes watering, breathing heavily and sweating profusely, I now had to put the suit on. As I was in the middle of a lane, I had to do this standing up, hopping around and trying to maintain some semblance of dignity and calm, while my inner child blubbed unashamedly and begged to go home. Whatever the pain though, I had to continue. I'd committed myself to becoming a beekeeper. I couldn't just give up because I'd been stung, even if it had been more than once and before I'd even got to see any bees. (In all honesty, I might have considered it if my beekeeping mentor hadn't been waiting patiently for me a few yards away, unaware of the life-and-death drama being performed just out of his sight.)

When they teach you the first principles of apiculture, there's a second Very Important Rule most beekeepers will impress upon you soon after the first one (You Will Get Stung, remember?). It's this: make sure yours is the only body inside the beesuit. Sounds obvious, I know. To be honest, there's not an awful lot of room for anyone else inside one, so on most occasions you don't have to worry. But every so often, a bee is going to try to get in there. And then things can get awkward. Still, it's quite rare, and if you take the right precautions it should never happen to you.

Of course there's an exception to every rule, and on this particular day my suit had an unwelcome visitor – and it wasn't just the guy tripping over his trousers in the car park. I was writhing, wriggling, and beginning to rather reconsider the whole project, but I determined to carry on. Having stepped into the trousers, I pulled the top over my shoulders and reached back for the veil. It comes over the top of the head like a hinged helmet, and zips round either side of the neck for a secure, supposedly bee-free fastening. My head was throbbing like an anvil, but I was at last ready to enter the apiary. The two bees that had stung me had disappeared, and were presumably dead. But now a third one appeared on my veil. It was quite calm, taking its time, walking around a bit and then stopping right in front of my eyes. I thought it was odd to have a bee for such close company, but then I focussed on it more closely and analysed the situation.

If that bee was on the outside of my veil, I should have been able to see its belly. So why, then, could I only see its back?

Like a child being taught its first lesson in arithmetic, I came to the correct conclusion slowly and after some prompting. 'So why has Little Miss Bee got her back to us? That's right, darling, it's because she's *inside* the veil.'

Now, there's getting stung on the head, which isn't nice, and there's getting stung on the face, which is a whole lot not nicer. Severe facial blemishes are not a great asset for a television presenter, and at that moment my whole career didn't so much flash before my eyes, as simply vanish. I could see the headline: TV NEWSMAN HIDEOUSLY DISFIGURED IN RAMPAGING BEE ATTACK. *Sting on tongue rules out radio as well.* So I did what so many characters in thrillers do in tense circumstances. I froze.

You can do a lot of thinking when you freeze.

Uh oh, there's a bee in my veil. Okay. Don't panic. There's just a bee inside the veil. OhmygodtheresabeeinmyveilwhatamIgoing todo?

Salvation, in the form of Christopher my bee mentor, was just a few yards away. All I had to do was walk over there, very slowly, and get his help. If I'd had a stick of nitroglycerine balanced on my head, I couldn't have stepped more carefully. It was just a few yards, but it felt like one of the longest walks of my life, played out to a disconcertingly loud soundtrack of buzzing. I must have looked rather strange, inching my way forward like an astronaut taking his first steps on the moon. But at last I made it through the gate and presented myself.

'Hello, I'm Bill Turnbull and I'm here to learn about beekeeping from you and, er, there's a bee inside my veil.'

Rather like John Wayne, Christopher stepped forward and reached out a gloved hand, found the intruder still perched a centimetre from my nose and pinched it. I almost expected him to drawl, 'The hell you say.' But of course he didn't, even when I added my pathetic punchline:

'And I've just been stung on the head. Twice.'

There are one or two possible morals to this story, lessons about life that you could usefully store away. And I know you may be thinking the first one is blindingly obvious: Stay Well Away From Bees and Beekeeping. But, by the time you've finished this book, I hope to have convinced you that you'd be very wrong to draw this conclusion. I certainly didn't realise it then, but beekeeping has been for me as fascinating, fulfilling and rewarding a pastime as any you could hope to find. Honestly. And with a bit of luck, in the following chapters you'll understand why.

On that first day at the apiary, I had already known that there was going to be a whole lot more to beekeeping than just turning on a honey tap at the front of the hive. I realised that there would be the occasional discomfort. But the episode with the swarm and the concrete post, and the ensuing shots of pain in my head, had taught me a valuable lesson: fortitude. That is, the power of firmness in the endurance of pain or adversity. I suppose I could have backed down, taken off the beesuit and driven home with a quick medicinal detour to the chemist. But I had to stick to it because other beekeepers were there and it would have been an embarrassing and rather pathetic admission of failure to pack it in on the very first outing, before even making it to the hive. Fortunately for me, fortitude came to my aid. Sometimes you've just got to stick to your guns and get on with it.

So despite everything, I look back to this first day at the apiary fondly. I like to think that the bees taught me something that day, and that I came away a slightly improved individual. It was the first of several life lessons they – and my fellow beekeepers – have taught me over the past few years, and was the moment I first realised that even by being a Bad Beekeeper, you can become a better person.

Chapter Two

ENTERPRISE

*In which we recall how we got started at this malarkey;
how even tough guys can be scared of bees; and our gratitude
to a dear departed chicken, to whose memory this
chapter is dedicated.*

I lived next door to a Hell's Angel once. Well, a retired Hell's Angel. His name was Richard and he didn't ride a motorbike any more because he'd broken virtually every bone in his body falling off one, and now he had kids. But he still seemed tough enough. Richard worked with his dad as a builder, and very good builders they were too. He read the *Guardian* and they listened to Radio Four while they were working. And they charged rates commensurate with people who read the *Guardian* and listened to Radio Four.

We lived side-by-side in two semi-detached cottages in a

small Buckinghamshire village. Richard and his wife Sue were great neighbours. One night when we had a chimney fire he came in, threw a wet sack on the grate and carried it out of the house. By the time the fire brigade arrived it was all over. The firemen were really disappointed.

Another time, just for fun, he planted our Christmas tree on the roof of the house.

Another time he hid behind a bush in our garden and, in the friendliest way, fired a shotgun. Can't remember why.

And, years before the *Lord of the Rings* films, he had a vicious Jack Russell (I mean, truly vicious) called Gollum.

As you might imagine, Richard was not a man to get on the wrong side of. It was a family trait. Although his forebears had lived in the village for hundreds of years, some of them had ended up on the other side of the world, transported there at His Majesty's Pleasure for a variety of offences, such as poaching deer or digging up someone else's potatoes. At one point, Richard himself had endured an unscheduled break in a Saudi Arabian prison when he was caught with some cannabis on a building job. So law-abiding was not foremost among his characteristics. If something got on his nerves, he did something about it – legally or otherwise, as I discovered one dark night.

Some neighbours a little way across the road from our happy homestead were in the habit of going away and leaving their dogs – two Spaniels – alone in the house. Someone would come and feed them, and they'd get in and out through the dog flap at their leisure.

One day though, something was wrong. One of the dogs was stuck outside, and yapping. For hours. Towards midnight I went over in the dark to try and sort it out. Once there, the problem was easy to remedy. A boot had fallen in front of the dog's door, stopping it from getting back in. Boot moved, dog re-housed, job done. Until I turned around. There was Richard right behind me. Carrying a knife. With intent.

If I'd been a minute later, there would have been one ex-Spaniel and a whole lot of explaining to do. It was like something out of *Straw Dogs*. Although if Richard had had his way, it would more likely have been Raw Dogs.

Not long after that, we moved. Not because of our dear neighbours, but because with three toddlers, the house was now too small. Just a few miles away, we found something bigger, something with a larger mortgage, and something that needed a lot of work. Perfect for a *Guardian*-reading, Radio Four-listening ex-Hell's Angel builder.

So why am I telling you all this? Well, I couldn't imagine that anything on earth could make my shootin', fishin', Spaniel-threatenin' ex-next-door neighbour Richard back off. Until one bright spring day in May, when he and a mate were outside in the garden working on the newly-bought-much-loved-but-in-need-of-some-restoration house or, as he put it, 'this decrepit pile of crap'. Suddenly, he came charging in, out of breath, like a horse that had been spooked. What could possibly have made this Tough Guy run?

Outside in the garden, marauding over their new territory, was a cloud of invaders.

A swarm of bees.

If it happened today, I'd be clapping my hands with joy and running out to greet them. But that day we did what most people do in such situations: we shut the windows and called the police.

Not much later, a man came along from the local bee-keepers' association to have a look. All he had with him was a cardboard box and a pair of secateurs. By now, the swarm was hanging in a cluster from the branch of a weeping pear tree at the bottom of the garden.

All he had to do was clip the branch off the tree and lower it gently into the cardboard box. Later, when the last stragglers had settled in, he came back, closed the lid, and took them away.

Richard never seemed quite so tough again.

It amazes me still, but I think I had reached my late thirties before this, my first proper encounter with bees. I'd never seen them close up before. In fact, I'd never really seen them before at all. And me a quasi-country boy (my parents had run a smallholding in Surrey). I'd seen wasps, of course. And been stung by them. But not bees.

So when the man with the cardboard box came to take away the swarm, I was truly impressed. There seemed to be something rather Zen-like about his ability to tame a large cluster of potentially ferocious flying insects and coax them into captivity. There had been no excitement, no fuss; just a sense of being at one with nature. In that moment, a flame was lit for me. Just a tiny little one. But as I watched, I wondered, could I have what it took to do what that beekeeper was doing? Did I have the skill, the calm, or the courage to manage such a situation?

I had to wait a long time for the answer, for work rather got in the way. Years passed, a decade and more. I travelled much of the world reporting for the BBC, covering elections, riots, uprisings, plane crashes, explosions, hurricanes and one very short South American war in which no one was actually hurt. As the next assignment was only ever a phone call away, I couldn't commit to the responsibility of looking after any livestock, let alone swarms of bees. The idea didn't so much get pushed onto the back burner as shoved almost right off it when we went to live in Washington for four years. Amid all the bullets, bombs and ballot boxes, I never once saw a bee, or even thought about them much, if at all.

It was thanks to a chicken called Tabasco that I eventually got involved. We'd started keeping hens in our kitchen garden when we came back from America, as it seemed a more profitable use of the space than just watching weeds grow. Just three to begin with. We let the children give them names, which is always a mistake, because of course they then become emotionally attached. If you have to wring a neck, as sometimes sadly you do, it's easier if you haven't been calling your feathered friend Dolly or Polly for the past three years. But here they were: Agatha, Sally (later to become known as Fat Sally when she stopped laying) and Tabasco. Why call a chicken Tabasco? Beats me, but that was the name bestowed upon her.

They were good hens, good layers, too. So good that in their first year they won first prize in the village horticultural society's annual show, for a Display of Three Eggs. I was very proud, even though it wasn't a big field; in fact mine had been the only entry. But as they say in football, a win is a win is a win. And we most definitely won.

As ever, these halcyon days were not to last. Tabasco fell ill, and a trip to the vet was called for. Out of interest, our daughter Flora, who was nine at the time, asked to come with me. Like many girls of her age, she was entertaining the idea of perhaps becoming a vet herself one day, even though she had once witnessed the forced evacuation of the dog's bowel by the insertion of a veterinary finger. 'Well, Daddy,' she had said philosophically, 'he was wearing gloves.' Anyway . . .

As we sat in the surgery with a cardboard box at our feet, waiting for Tabasco's one and only visit to the hen doctor, Flora's eyes wandered around the room, until they settled on an intriguing notice.

'Look, Dad,' she said. 'Beekeeping classes.'

And there it was, an invitation from a nearby beekeepers'

association for aspiring apiarists. It was clearly a sign. It was meant to be.

As moments of inspiration go, I'll admit it wasn't quite up there with the greats. Archimedes had his bath. Robert the Bruce had his spider. And I had the notice in the vet's waiting room. Still, it was enough. I decided that the moment had arrived to take on the challenge. And I have to say, I've never regretted it.

Sadly, things didn't look so bright for Tabasco. She was suffering from a form of peritonitis, and the prognosis was not good. It seemed she was destined for the great chicken coop in the sky. The vet offered to sort her out for the usual fee, but I wasn't brought up on a smallholding for nothing, and we went home with heads hung low, to commit the dirty deed.

Tabasco's legacy endures, though. Without her, you wouldn't be reading this now. For not much later, I signed up to be a new pupil with the Pinner & Ruislip Beekeepers Association, and duly turned up with my two fellow students every Thursday night for my induction course. We learned about the life cycle of the bee, how the hive works, the different types of equipment and, most importantly, how to knock some of it together. Although we were all eager to get our first hands-on experience, Tony – our tutor – wouldn't let us near a live bee until we'd mastered the basics. And quite right he was too. Beekeeping is not something to be rushed into. It requires patience, application and responsibility.

Somehow I managed to convince him that I had all three. If only he'd known . . .

That day, all those years ago, when the bees landed on the weeping pear tree, I never imagined that I might gain the expertise myself to go along to someone's garden with a cardboard box and take away a swarm. If it hadn't been for the chicken, or for Flora spotting the sign, I might never have done so. And above all, if I hadn't taken the opportunity it wouldn't have happened. Seize the day. *Carpe diem*. Although what fish have got to do with beekeeping is beyond me.

Chapter Three

CIVILISATION

Wherein we examine the inner
workings of the bees' domestic arrangements;
wonder at their discipline and organisation; and ask
ourselves: if they're so clever, how come they're
not ruling the world?

So how does it all work? First off, I'm not an expert. You knew that already. There are any number of very good books on beekeeping theory which can teach you the science of it all. But you're not going to read those. Not just yet. And if I wrote it down it would sound like a biology lesson. And as I just scraped a pass in my Biology O-level (yes, we are going back a long way) I wouldn't recommend that. So let's do what I always do when talking to people about beekeeping. You ask me questions, and I'll give you answers. Go on. Ask me a question. Not that one, a proper question. About bees and stuff. Okay, here are the things I get asked the most.

Aren't you worried about getting stung? Doesn't it hurt?

Yes it does, but that's where the fun is. If bees didn't sting, it would be a bit like keeping flies. If there's no risk, there can be no adventure. And they only sting for a purpose; if they're feeling threatened. Usually. I must confess I've suffered a fair few stings in my time. But on most occasions I deserved it.

How do bees make honey?

The Bad Beekeeper's answer: Er, well, they, like, go out and find flowers and bring this nectar stuff back to the hive and then, like, sort of make honey out of it.

The Good Beekeeper's answer: Bees make honey from the nectar that they draw from flowers. They extract the nectar – a sugary liquid – out of the flowers and store it in one of their two stomachs. They take it back to the hive and regurgitate it onto other bees. Then they store it in the honeycomb inside the hive, where the water content evaporates. The bees also fan it with their wings to make it dryer. When it's ready, they seal the honeycomb with a thin layer of wax.

What's the smoker for?

Good Beekeeper: The smoke makes the bees think that there's a fire nearby and that they may have to evacuate the hive. They quickly eat some of their honey to prepare for the journey. When the smoke passes, they realise that they won't have to leave after all. By this time though, they've had so much to eat that their mood has improved and so they are more placid. There is another theory that when the bees' stomachs are full, they can't bend their bodies enough to sting you.

Bad Beekeeper: The smoke gets in their eyes and then they can't see you.

How does the colony actually work?

Bad Beekeeper: Well, there's this queen, see, and she rules the roost and lays loads of eggs and the other bees all protect her: the workers do 'cos they're all women, and the others – the drones – are men, and they just hang around all summer eating honey and hoping to make it with a queen. Only it's a bit of a bummer when they do, 'cos their knackers snap off and they die.

Good Beekeeper: Each hive is a city in itself, with its own very clear rules and ways of doing things, and a rigid system of self-government. Bees can't talk to each other, but they do communicate through smell and movement. They can tell each other where the best sources of food are, and how far away it is. But somehow they also have a collective consciousness that determines their lives. Their society is layered in several different strata. The moment they are born, they are given a task to complete – cleaning out the cell from which they have just emerged. Thereafter their lives consist of several different jobs. They start off as house bees: maintaining the order of the hive, cleaning up, dragging the bodies of dead bees down to the entrance, and feeding the larvae that have yet to be born.

As they get older, they actually make the honey. Some of them process the nectar that the foragers bring in. They pass it from mouth to mouth, adding enzymes, before depositing it into the cells for storage, and ultimately sealing it with wax when it's ready. So the honey you eat has actually been through the mouths of a number of bees, and been expectorated from one to another. Yum.

The next rung up the bee career-ladder is building. The bees fashion new cells out of wax, which comes from glands on the underside of their abdomen. Left to their own devices, the girls will fill any vacant space with wax comb; it is beautiful to behold – thousands of white hexagonal cells, all the same size, all perfectly formed.

But the senior job inside the hive is security, reserved for mature bees whose stings are fully developed and loaded with a good amount of venom. If you look into the entrance of a hive you will often see a few of them standing just inside, hanging back, ready to deal with an intruder. Rather like bouncers. Any bee which lands in there will be challenged by the guard. If it has the right pheromone – if it smells right – they'll let it in. If it doesn't, they may allow her to pass anyway after a brief negotiation, especially if she's carrying nectar. Just like at a nightclub:

'Sorry, love, you can't come in, you don't smell right. This is an Armani hive.'

'Yes, but I'm wearing Ralph Lauren and I've got a nice little bit of sweet stuff here, wanna taste?'

'Oh, all right then. Come on in.'

There are exceptions to this later in the summer when the robbing season is under way. That's when there are no more flowers to be worked, but the sun is still shining and the bees are still flying. If they smell honey and come across another hive, they'll try their luck and may on occasion mount a raid *en masse*. With so much at stake, the guards can be particularly aggressive at this time. Just try shoving a stick in the entrance and see what happens.

The elite of the colony are the foragers, the only bees that fly regularly. They're the ones who keep the colony alive by going out and getting the food that the community needs to survive. So as soon as they feel the heat of the sun on the roof, they go out and look for supplies. Some have the job of bringing back pollen, which is their source of protein. You can see them waddling back into the hive with little drop-sized bags of pollen on their back legs. And you can tell which sort of flowers they've come from by the colour of the pollen they're carrying. So, orange may be dandelion; clover produces brown pollen.

It's always a good sign to see the first pollen being brought back in spring, because it usually means there is a queen in the colony laying eggs. If there's no queen, or she isn't laying, the bees don't bother with the pollen and you'll know you may have a problem without even having to open up the hive.

Some of the foragers are water-gatherers, and if you have a little dribble of water somewhere in your garden, you may see bees coming to it and taking a drink, not for their own satisfaction but for the benefit of the hive – they take it back to the colony for further distribution.

But it's the nectar that gets all the attention from us, of course, because that's what makes the honey. Each flower in the garden, or in the wild, contains at its heart tiny amounts of a sugary liquid. The bees suck it out with a long tube-like tongue, and then store it in one of their stomachs. The foragers work incredibly hard. From dawn to dusk and beyond, they're out working on the flowers, sometimes up to two miles away from their home. They work out where they're going from the position of the sun. And when it's cloudy, they can still calculate where it is, by using polarized light.

If you watch bees coming out of the hive, they'll often fly up towards the sun initially in order to get their bearings, before veering off in a particular direction. That's because they've been told by other bees where to go. They do this inside the hive by performing what's called a waggle dance. First of all they march up and down a bit to indicate the position of the sun. Then they veer off at an angle to indicate the compass bearing of the flowers they are guiding other bees to. They also shake their bodies at a certain frequency to let the others know just how far they have to go. And here's the really interesting bit: the other bees can't see them doing it. They *feel* the dance through their antennae. It's astounding when you think about it.

In its entire life, a single bee gathers enough nectar to make one teaspoonful of honey. Just the one. To produce a pound of honey, it's estimated that bees collectively fly about fifty-five thousand miles. In the summer, bees only live for six weeks. And it's only in the last third of their life that they get to go out and do what everyone expects them to do.

There's a good reason for this. At a certain level of bee government, they worked out that it is more efficient for a bee to end its life outside the hive than in. If a bee dies at home, it means that others have to spend valuable time dragging it down the various layers of the hive, through the frames to the entrance, and then chucking it out. And have you ever tried removing a body from the sixth floor of a tower block?

On the other hand, if the bee dies while out on the wing,

there's no cleaning up to do. She's just gone. And when she doesn't come back, no one will miss her. They probably won't even register her absence. It's a brilliant system. The spirit of the beehive governs all.

So much for the women. But there is – there has to be – a male component, in the form of drones. These are the boys, and sadly they do rather give their gender a bad name. They are, compared with the workers, larger and rather fat looking, with big round bottoms. They are lazy and indolent. They bring nothing into the community in terms of food. They make a lot of noise when they fly, probably because they're so tubby it takes much more effort to get them airborne. And they've got just one aim in life: yes, that one. All in all then, they're pretty much like the rest of us boys.

The problem for them is, though, that there's not a lot of talent out there, to put it mildly. The working females don't mate, so that just leaves the queen. There's only one in every colony, and she's only up for it once in her entire life. So the opportunities for a keen young drone are desperately few and very far between.

It should be said that although the queen only goes out on a few mating flights over a couple of days, she does have quite a time of it. She takes off from the hive, and the drones are drawn to her by her scent. Over the next few minutes she mates with not one but up to twenty different drones, building up a store of semen which will last her for the rest of her life – which could be several years. So a number of drones will fancy their chances, which is why a cloud of them will appear around the queen. But here, for them, is a fatal catch. Once they have mated successfully, their genitalia snap off. And they die. You can imagine the conversation taking place in the hive:

'Oi, Bert, you know that cute little buzzer they've making all that fuss about?'

'What, the one next door that acts all stuck up like a princess?'

'Yes. Well, she is the queen after all. Anyway, I hear Her Majesty's in the mood, so to speak.'

'Yeah, can you smell what I smell? Phwoar. That's irresistible.'

'Gerraloadofthat. I'd give anything to have a go with her.'

'Yep. She's to die for, all right.'

The probability of a drone doing what he's born to do is pretty slim. In any colony there are hundreds of drones and just one queen, and chances are she's already mated and could live for up to five years. So, the drone has to find a newly born queen who's just emerged from a hive close to where he lives, or possibly in his own colony if the old queen has swarmed. Then he has to get to her before all the other likely lads. The odds of a happy but fatal outcome must be hundreds to one, if not higher.

So, most drones spend their lives whiling away the

summer days eating honey, taking up space inside the hive, and probably giving the hard-working females a bit of backchat into the bargain.

Until late August that is, when the worm turns in a dramatic fashion. The workers, sensing that the summer is coming to an end and the time of plenty with it, realise that those fat and lazy layabouts who've been hanging around all summer are actually serving no useful purpose any more, and will only consume valuable supplies if they survive into the winter. So they start a clearout. They either evict them by pushing them out of the hive, where they will have to fend for themselves, or they take the more direct route and kill them. The drones are defenceless, for they have no sting. It's a brutal but highly efficient operation.

What amazes me about all this is the way that everything is worked out. It's a very cleverly organised system, perfected over tens of millions of years. Everyone has their job. Everyone knows their place. And it's all done, apparently, by instinct. There is no formal government, no top dog as such, and certainly no alpha male. You might expect the queen to be in charge of course, but it's a tough one to argue. True, she is fussed over and cosseted as the colony's most precious asset, the one member on whom the future of the community depends. But she has no governing status. She is a breeding machine, laying up to 2,500 eggs a day at the peak of the season. Once she has mated, she does not leave the hive again, except in a swarm to find a new home and make way for a new queen behind her. She stays in the warm darkness of the hive. For years.

So, could we learn anything from the way the bees live? Well, it's crazy to try to anthropomorphise insects – to equate their way of life with our human existence – because their biology is so very different to ours. And, more importantly, you could argue that they're actually much more civilised than we are.

Like other insects, their lives are centred around the community, the greater good. There is no selfishness, no greed. The individual sacrifices herself willingly for the sake of the sisterhood. All that matters is the survival of the colony; everything else – *everyone* else – is subordinate.

And they never sleep. And they work till they drop.

Sounds awful, doesn't it?

HOW A HIVE
WORKS

Let's start at the bottom. First you've got your stand.

It can be right down on the ground, or a couple of feet high for ease of access.

On that, we place the floor.

Traditionally this is made of wood. More often these days we're using a fine wire mesh to combat the evil varroa mite.

Above that, the brood box.

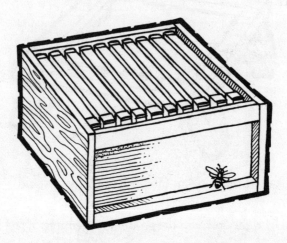

This is the heart of the hive; the engine room. The queen lives here along with most of the colony, and this is where the eggs are laid. They turn into larvae and then pupae, and eventually emerge as fully formed and full-sized bees.

The brood box, and other boxes, are occupied by wooden frames.

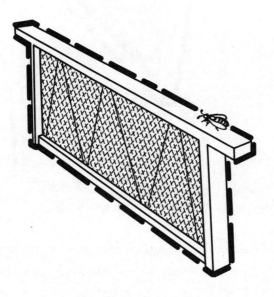

They are literally just frames holding a sheet of wax foundation, which the bees chew and draw out to make the cells. The cells are used to incubate eggs and larvae, and to store pollen and, of course, honey.

On top of the brood box is placed the queen excluder.

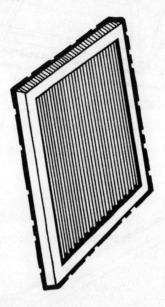

This is a metal or plastic grid, which allows access for the smaller worker bees, but which the queen cannot get through. It's useful because otherwise she would be laying eggs all over the hive, and that would mess up the honey.

Next comes the part most beekeepers live for, the *super*.

This is like a brood box, only shallower. This is where the bees store their excess honey. We like that. You can have as many supers as you like; though of course the taller they get, the harder they are to take off later.

On top of all that you put the crown board; just a wooden ceiling really, and then the roof. And the whole thing looks like this.

Chapter Four

LABOUR OR HOW
WE MAKE IT WORK

*We learn how not to make honey. The magical process
of extracting and bottling. And how a dog is not always
a beekeeper's best friend.*

So, we know how wonderfully clever and everything the bees
are now. But there is one flaw in their otherwise perfect but
sometimes rather unsettling society. For some reason, at some
point, they decided it was a good idea to produce a surplus
of honey.

Why is it that we're all beekeepers and not bumblebee
keepers? The bumbles, fascinating in their own right, don't
produce a surplus. They have small nests and only make and
keep as much honey as they need. Honeybees, on the other
hand . . . well, if only they'd known.

I thought I'd bring you to the sharp end of beekeeping first. There's lots of other stuff about maintaining hives in good order, making frames, preventing disease, and all sorts of exciting procedures such as the shook swarm and the use of the Cholmondeley-Smithson clearing board. But you really want to know how we make honey, don't you?

Well, *if* your bees haven't swarmed and gone away and left you with half a workforce, and *if* it hasn't rained for most of the summer so that they've already eaten everything they'd produced, and *if* there hasn't been a drought stopping the best flowers from blossoming, hence depriving the bees of the nectar, well *then* you might actually get to harvest some honey.

You might think that this should happen every year, and it's true that most of the time there are beekeepers in some parts of Britain who've been able to fill a few jars. But you'd also be amazed at how weather-sensitive we all are. The summer of 2008 was a disaster for beekeeping in the UK. First of all, large numbers of bees had died in the previous winter, giving rise to all sorts of fears about Colony Collapse Disorder and the death of the bee and the end of the world, most of which proved to be unfounded; more of which later. The real horror was that the summer was so wet, and it rained so persistently, that very little honey was produced.

When the sun shines, the bees fly out and gather the nectar to make the honey. But when it rains, not only do they all stay at home, but they sit inside eating up all the honey they've produced, so it becomes a double negative. With week upon week of rain in the summer of 2008, by December there was no English honey to be had in the shops. None.

And my crop that year? Nil. Nought. Nothing. Let's move on.

But let's make believe for a moment though, and imagine that we've had a glorious British summer. Crazy, I know. But we can dream. So the sun has shone from dawn to dusk, and bees have flown merrily to and from the hives all day long, and then as darkness descends every day, a little shower falls to water the flowers so that they don't dry up, and all is well with the world. It's a fantasy, remember.

In this ideal scenario, the bees first of all stack up all the honey they need in the brood box at the bottom of the hive. Of the eleven frames in there, the four outer ones will be all honey, and the inner ones will have mostly eggs and larvae, with honey round the outside. So, if this box is full, anything above it is superfluous. Which is why we call the other boxes in the hive 'supers'. Clever, isn't it?

Actually, I'm just guessing. The reason they're called supers is because that's what beekeepers say when they find they've got some honey in there.

No really, the real reason they're called supers is from the Latin preposition *super*, meaning over or above. As the super is above the brood box, that's what it's called.

Anyway . . . once you find a super full of honey that has

been sealed, you can take it off. Some beekeepers can do that in the spring if they get what's called an 'early flow' from crops like rape seed. But the biggest harvest comes in July, when the main summer flow comes to an end. At least that's when the efficient beekeepers take their honey off. The rest of us do it in August. Or September. Or, erm, October. In terms of the honey itself, it doesn't really matter. If it's sealed, it'll keep, although there's a fair chance that the bees will consume a lot of it once the colder, wetter weather sets in.

Once you've cleared the bees out of the super, which can be a hazardous experience in itself (more on that later), you can lift it away and carry it to the car. Sounds simple enough in principle, doesn't it? Remember, though, that it is probably still the height of summer. The sun is shining. It's a hot day. The super weighs somewhere in the region of twenty-five to thirty pounds. The car is some distance from the hive. And you're wearing a beesuit. Zipped up to the gills, with the veil on. You know the expression 'bathed in sweat'? I'm sure it must have been invented by a beekeeper, because that's what it feels like. And you can't wipe it off because you've got your suit on. And you've got your veil on because there's a fair chance that you're accompanied by a number of bees who may be curious, to put it mildly, about why you're taking half of their house and all the food it contains away from them.

I love it when the sweat is drizzling into my eyes. It has a really sweet sting to it. And as you trudge blindly towards the car, you wonder if you left the boot open or closed . . . And if it's closed, did you lock it? And if you locked it, where are the keys?

The thing is, on your first trip to the hive, you can leave the back of the car open in anticipation of your return. But once you've loaded one, you rather want to shut the boot to stop more bees getting in. They have this pesky habit of trying to claim back what is rightfully theirs. So with supers number two and beyond, you arrive at the car carrying your thirty-pound box and try to hold it on your left hip, while with your right hand you attempt to open the boot. But first you have to shake off the glove you've been wearing, because otherwise your fingers are too fat and clumsy to push the button. Of course, if the boot happens to be locked, then you have to fiddle around in your right-hand pocket feeling for the keys, then remember that the keys are in the *left* pocket. And so begins the rather awkward contortion of reaching round for the left pocket while at the same time holding the thirty-pound box on the left hip. Or transferring the thirty-pound box to the right hip and searching for the keys with the left hand. Oh, but there's a glove on it . . .

I forgot to mention that by this time there will be a fair amount of honey around, as large dollops of it come away when you separate the supers to lift them off. Rather inconsiderately, the bees don't take convenience of removal into account when packing honey into every available nook and cranny in the hive. You can reasonably expect that by this time, your gloves will be quite well oiled with sweet, sticky stuff. The gloves you are now trying to remove. Or the gloves that you are now manically dipping into your pocket in a hurried search for the car keys.

And no, you can't put the super on the ground to make things easier. It's dripping with honey, isn't it?

So at last you get the car open, and now you can put the super in there. The super dripping with honey. You did line the boot with plenty of newspaper, didn't you? Didn't you? Oh dear. Never mind, honey washes away quite easily with a bit of soap and water. It's the propolis that you can't get out. What's propolis, you ask? That's the even stickier substance that the bees have laid round the edges of the super to keep the hive glued together. It's the stuff that is now sticking the super to the carpet of the boot where you didn't put the newspaper down.

But now you've taken off all the honey, and put the lid back on the hive, and you can leave. You still have to drive with the veil on, for you have an escort of several dozen by now rather angry bees. The thing to do here is to drive for a bit, then stop after half a mile and open the boot to give the bees the chance to get out, for by now they will want to go home. It's a good idea, and it does work. Just don't let them out at a motorway service station. And no, even *I* haven't done that.

It's fair to say that harvest time is the occasion in the bee-keeping calendar which tests marriages the most. For the rest of the year, the beekeeper's spouse can get on with their life unbothered by the goings-on in the apiary. They will nod and murmur assuredly when the beekeeper talks to them about the latest complications involving the varroa mite or outbreaks of nosema, even though they really don't understand what they're on about. They will sympathise with, and tend to, the beekeeper's wounds: 'You've been stung *where*?' They know that beekeeping is a generally constructive pastime which is keeping their partner (a) busy, (b) reasonably content and (c) out of trouble. And it's not costing very much. They will be comforted by the fact that, even if the beekeeper comes home three hours later than promised, they will know where he's been.

And it could be a lot worse. When, on the very odd occasion that Mrs T questions the amount of time I spend at the hives, I need say only this:

'I could be playing golf, you know.' M'lud, the defence rests.

And from time to time, they do reap the dividend of a highly prized jar of honey, although it's those jars that can really do the damage.

Let's go back to the sweat-soaked but merry beekeeper who arrives home with his precious and hard-won load. Puffing and panting, he appears at the kitchen door, ready to begin the process of extraction. There is not a second to lose, particularly because the super he is holding is dripping honey onto the floor. Space must be cleared and surfaces must be wiped, for great work is about to be done.

What he has to do first, when he has deposited the sticky super next to the sink, is get the extractor in. This is a large cylindrical drum on three legs, which is used to spin the honey out of the frames. For the past twelve months it's been down in the shed, gathering dust and possibly the attention of the local wasp population, which has cleaned out every last drop of sweetness that may have been lingering there.

It is a large and cumbersome device, and in bringing it into the house the Bad Beekeeper may happen to bump into the doorposts once or twice, chipping the paintwork and cranking up the spousal tension by a further notch.

Once the extractor is in the kitchen, he realises that it needs to be washed clean. Then he remembers that it is far too big to fit into the sink, and that cleaning it in the kitchen does slop rather a lot of water onto the floor; this is also a tension-cranker. So, the extractor has to be taken outside again, with more door bumping and paint chipping, where it can be properly washed and rinsed. More time is spent waiting for it to dry, for the one thing that is the death of honey is moisture. It makes it ferment, with ruinous consequences. And, of course, during that time, honey from the sticky super is now sliding over the front of the draining surface and down the kitchen cupboards, towards the floor . . .

Bad news travels fast, they say, but not quite as fast as honey on a kitchen floor; the tiniest drop, once stepped on, can be distributed all over the house in a matter of seconds by the right pair of feet. Like those of the dog.

So you see, what just a few minutes ago was the very picture of good order and domestic bliss has now become a battlefield. On such occasions it would probably be a good idea to suggest that one's spouse spend the day away somewhere less aggravating, like at a Hell's Angels convention. Except that the Bad Beekeeper does rather need a spare pair of hands. And there's no way she is going to let him have the run of her kitchen unsupervised. It would be like letting a gang of ten-year-old boys run amok trying to make chocolate cakes. Only messier.

Now, the extractor comes back in (more bumping, more chipping), attended by a number of bees who have quickly picked up the scent of honey, however faint, emanating from the tin drum drying in the garden. At this time of year, with no flowering plants to explore, they quickly latch on to anything that might provide a source of food. With a vengeance.

Undaunted, the Bad Beekeeper can start his work proper. The first job is uncapping – taking the wax seal off the frames of honey – otherwise the honey will not emerge from the cells in the extractor. It's like taking the lid off a tin of beans. If you don't take the top off, you can't get to the beans.

There are several ways to uncap, each with its own foibles and disadvantages. Most beekeepers start with the uncapping fork. This is usually a little plastic handle with a dozen sharp steel prongs on the end. You take the frame of honey in your left hand, support it on a firm surface, and slowly skewer across the tops of the cells, so that the wax capping comes off and the honey can be released. It's quite a good method, but rather laborious. The handle quickly becomes very sticky, the wax soon builds up on the prongs and has to be scraped off. This, for the Bad Beekeeper, means lots of stops to attend to puncture wounds.

Next up is the hot air method; not *taking* the cappings off the honey, but *blowing* them off. You get one of those wall-paper-removing devices from a DIY store – like a hairdryer but much hotter and more powerful. Hot enough to take your hair right off. Applied gently to the frames, it magically makes the wax cappings disappear; they literally melt away. This is a good thing. The drawback is that unless applied very carefully, the heat also melts the entire wax cellwork holding the honey in place. So if you're unlucky, or inattentive, the whole structure can dissolve into a gooey mess.

So then we come to my personal favourite. The sword. Actually it's just a heated blade, not a sword. But it looks like one and it reminds me of the knights of old, so as far as I'm concerned, it's a sword. It has an electric element inside it, so once it's plugged in, you can carve right across the surface of the cappings to reveal the honey underneath. It's also good fun, so long as you take great care where you put it down. As the temperature of the blade can reach 1,000 degrees (I'm guessing), you can easily end up with a scorched work surface (oops), or worse still, a scorched hand.

Once each frame has been uncapped, it can be placed in the extractor, ready for spinning. This gives the Bad Beekeeper time to clear up some of the mess he's already created. The cappings, along with a generous coating of honey of course, are supposed to drip sensibly into a dish underneath the frame. Naturally, during various puncturing and scorching moments, control of the proceedings may have been lost. And some of the material may have ended up outside the dish, on the work surfaces, in the sink, or on the floor. Where the dog is licking it up. Who let her back in?

Sooner or later – when he's beaten off the wasps who have also tracked down the gorgeous smell of freshly released honey and are buzzing contentedly around his face – sooner or later, when he has uncapped enough frames, the Bad Beekeeper can get the extractor going. It's one of my favourite bits of beekeeping gear – big, sturdy, powerful and full of promise. In fact, I like it so much it once stayed in our kitchen/dining room for a full year, standing majestic in the corner like a priceless Ming vase. Mrs T was horrified at the prospect of entertaining guests in the presence of such a monstrosity, and requested firmly, more than once, that it be removed. But it had to stay where it was for important technical reasons; principally, that it still had honey in it. Quite why it hadn't been put into jars, I can't remember now. But something had come between me and the bottling process that year, and the honey resting in the extractor had solidified, as it often does in winter. It could not be tapped off, and so the great machine – known affectionately as the Beast – had to stay in the house.

'Anyway,' I said in a vain attempt at spousal pacification, 'it'll be a talking point.'

And it certainly was. Through Christmas and New Year parties, Sunday lunches and even my fiftieth birthday dinner, many a sparkling conversation opened with the words, 'What the hell is that?'

It was a question some guests came to regret, as it allowed me to launch into my 'Why Beekeeping is So Fascinating' lecture, which never lasts less than five minutes. (Think how long it's taken you to read this far, and you'll see what I mean.) When their eyes began to glaze over, I wondered, if I'd been too generous with the *aperitifs*. But no, it was merely the hypnotic effect of my treatise on the benefits of apiculture lulling them into a state of silent passivity. In other words, they were bored rigid. They always sprang back to life though, when I lifted the lid and invited them to take a sniff. Honey always smells good, especially in a tank. And they were probably delighted that the lecture was over.

So anyway, when the frames have been uncapped, the Bad Beekeeper can at last fire up the extractor. In terms of what's on offer, they're a bit like cars. They vary in size, the smallest taking just two frames, the largest a dozen. Some, called radials, have the frames slotted in like the spokes on a wheel. Others, tangentials, have them mounted facing outwards, like the rim of the wheel itself. Each has its own advantage, which is too boring to go into here.

Some have electric motors, but most just have a crank handle on top which you wind around. So, mine is a nine-frame radial manual. French. Bit like a Peugeot. Efficient, but no great sense of style or comfort. I hanker for an electric machine, but don't ever get enough honey to justify the cost. Such are the hopes and dreams of a beekeeper: 'Dear santa plese culd i hav a 12 frame lectric xtractor wiv enuf huney to put init i've been very gud reelly thankyou yors sinsearly bill.'

At last then, the Bad Beekeeper can begin what is, for me at least, the most thrilling part of the beekeeping experience. So long as he has the nuts and bolts to get it to work. Literally. For when cleaning the extractor after last year's harvest, he will have taken off the nuts and bolts that keep the lid on and the internal mechanism together, and will have put them somewhere for safe keeping. Now, where was that place?

I do have a bit of form on this one. I find hiding places so safe that no one ever finds them, even after searching high and low throughout the house. As his A-levels approached a few years ago, Henry, our oldest son, begged me to take away his Pro-Football Manager game, which was the joy of his life, but which rather got in the way of revision. Naturally, I was more than happy to help, and I took the far-too-tempting disc off him. It has, since then, not been seen again. By any of us.

So when the extractor's arrival approaches, it can be an uncomfortable moment for the Bad Beekeeper, even disturbing the bliss of a romantic holiday abroad:

BB: I think when we get back it would be a good time to take off some of the honey.

Helpful Spouse: That's a great idea. I'll help you.

Troubled look clouds BB's vision.

HS: What's the matter?

BB: Nothing, really. Just trying to remember where the extractor's bits are.

HS: You mean you didn't put them back on last year?

BB: Year before, actually. We didn't have any honey last year.

HS: But you didn't put them back on the last time you put the extractor away. So they've been missing for two years.

BB: Well yes, but not missing exactly. Just . . . resting. Somewhere.

HS: You've lost them, haven't you?

BB: I'm sure they're in the kitchen drawer. Or the shed. Or—

Knowing look from HS silences BB. They both know that he doesn't know where the bits are.

Pause.

BB: Cup of tea, darling?

Let's assume, for the Bad Beekeeper's sake, that the bits have now been found at the back of the kitchen drawer. The machine is properly assembled. The frames are inserted, the lid is closed; he begins to turn the handle, which drives the shaft, which spins the frames around, which flings the honey out of the cells onto the walls of the extractor, where it slides down the sides, through a filter which catches the extraneous lumps of wax, and into the tank below.

You're not supposed to lift the lid while spinning the frames. If you do, there's a chance that you might be tempted to dip your hand in to make an adjustment of some kind. And that would be a bad idea. You'd break your arm at the very least, and with a bit of luck you might even shear your whole hand off. This would be inconvenient because (a) it would require a trip to A & E and several hours, if not days, wasted, and (b) all that bone, blood and gristle would contaminate the honey.

But if you were to open the lid while you're spinning the frames, you would see millions of tiny droplets of honey spraying out of the frames like the finest of sprinklers. As it's caught in the light, it looks like a magic spider is spinning threads of silver and casting them on the walls of the drum. And a rich gentle breeze fans up with the most delicious aroma that says: yes, it's all been worth it.

It's usually at a moment like this that the Bad Beekeeper is woken from his reverie by a screeching. It can be one of two things . . . or quite probably both. The first screeching is coming from below, where the extractor's feet are carving wild graffiti on the floor of the kitchen. Wood, linoleum, tiles – you name it, the extractor doesn't care. Powered by the massive centrifugal force of the crank handle, it can carve its mark indelibly into any surface. It's probably something to do with action and reaction being equal and opposite; one of the first laws of physics.

Certainly, one of the first laws of marriage is that a wife, on seeing her beautiful kitchen floor being vandalised by her husband, will quite reasonably hit the roof. That's where the other screeching has been coming from. The Bad Beekeeper has forgotten to place the extractor on a wooden board so that it can slide merrily across the kitchen without causing any damage. Oh well, he can try that again next year. If he's still in the house.

Once the spray of honey has subsided, the empty frames can be taken out and replaced with a new set. What happens next is really important, for he now has what are called wet frames, which don't have enough honey left to harvest, but which are still very sticky. The good beekeeper conscientiously takes them to the apiary and gives them back to the bees, who will then clean the frames out so that they're spick and span for filling up again next year.

The Bad Beekeeper really wants to do this, but other things just get in the way. Like the new football season. Or the appointment with the marriage counsellor. So, the wet frames are taken down to the shed, where they attract the attention of first the bees, and then millions of wasps . . .

thus making it impossible to return the extractor to its proper home for several more months.

Back in the kitchen, the harvesting process continues. For there are still a couple of stages to be completed before he can relax and enjoy the fruits of his labours. The honey in the bottom of the extractor has to be tapped off into another drum, called a settling tank, so that it can rest for a couple of days and any air bubbles formed can slowly make their way to the top. The transfer from extractor to settling tank does, of course, provide the Bad Beekeeper with another opportunity to exercise his inability.

Naturally, in order to pour from one container to the other, the extractor must be higher than the settling tank, thanks to the laws of gravity. This means that it has to be lifted from the floor to a table, or work surface, strong enough to take the weight.

Ah, the weight. In his eagerness to process all the frames as quickly as possible, the Bad Beekeeper has forgotten the generous proportions of his nine-frame radial manual. It will take up to two hundred pounds of honey in its bottom receptacle, which is really useful when extracting, but makes lifting a challenge. Two hundred pounds. More than fourteen stone. Or ninety kilos, if you're younger than forty-five. And he's got to lift it onto the table. Plainly, that's not going to work, even if he could persuade his probably still-simmering spouse to help him. So he'll have to find some smaller receptacles so that he can tap off enough honey to make the extractor, eventually, light enough to lift . . . and he'll probably have to do that on his own. For by now the simmering spouse will be getting out the Yellow Pages and looking under D for divorce.

Once the honey is decanted into a series of saucepans with lids on – to keep the wasps out – and the attendant spillages have been wiped up as far as the eye can see (although the simmering spouse will somehow always find a few more), then the extractor can at last be lifted onto the table/work surface/kitchen island. Oh, we can only hope he remembered to put newspaper underneath it, to catch the drops . . . And something soft to cushion the weight so that the bottom rim of the extractor doesn't dig into the wood and mark it permanently. Never mind if he forgot, in future years it'll be a talking point for his partner, who'll remember the incident fondly: 'And that's where my ex put the honey tank.'

But when he's done all that, and cleaned up the mess again, the next bit is simple but fun. You just pour what is, by now, several gallons of honey through the tap at the bottom of the extractor, down through a sieve and into the settling tank. For the first time, before your very eyes, you get to see close-up what your bees have produced. It is a gorgeous, golden waterfall; thick and rich in colour and consistency, cascading in slow motion before you. And as the Bad Beekeeper watches it pour, he can wonder at the magic of this whole process; how all this has come from the simple blossoms of the flowers in his garden, and the extraordinary alchemy performed by the bees in making it. He can offer them a few silent words of thanks, and pause to meditate on the wonder of it all. Or he can just work out how many jars he's going to need.

There's another crucial piece of equipment here: the sieve. It might not sound so important, but you realise how useful it is once you can't find it. It's no ordinary sieve, you see. It's a German-manufactured stainless steel *doppelsieb*. It has two strainers: one flat and coarse, to stop all the lumps, and one with a fine mesh, shaped like a bowl. The flat coarse one fits into the fine one and they work very well together. Together being the operative word. One without the other only does half the job.

In one recent harvest, I could not find the top half of my *doppelsieb* anywhere. I searched high and low. In the shed, the garage, the kitchen cupboards. But it had vanished. No matter, I would just have to struggle on without it. There were other things to be done first, like emptying the settling tank of the honey that had been there for the past two years. And why, you might well ask, had it been there that long? Well, it was ivy honey. Ivy is the bane of the beekeeper's autumn. If the sun is still shining in September, the bees bring it in by the bucket-load. This would be a good thing but for the fact that it sets rock hard very quickly, and has an unpleasant, bitter aftertaste. When you first put it on your tongue it seems quite sweet, until it reaches the side of your palate and you wonder why your honey has been tainted with tar soap. I'm told it's really popular in France.

Anyway, two years previously the bees had generously given me a load of it. In an uncharacteristic flash of not-too-bad-beekeeping, I managed to get it out of the frames and into the settling tank before it could set. And then, with all the distractions of modern life, it was forgotten about. The seasons passed twice over, until the day I needed the settling tank again and remembered that it had had something settling there for two whole years. By now the ivy honey inside it was like concrete, though it did taste a bit better. So I had to warm it up to make it liquid again. And once reverted to its fluid form, it revealed its secret.

There, perfectly preserved at the bottom, was the top half of my *doppelsieb*. Someone, somehow, had dropped it in there. And forgotten about it.

After the last drop of his harvest is *doppel* sieved and poured, the Bad Beekeeper can put everything away for a few days, do his best to make things tidy, and see what can be done to patch up his marriage. When the dust has settled, it will be time for the last step: the bottling. This is the hour of greatest pride. He knows how a master vintner must feel when he taps off the first drops of his finest cognac. It is a dignified, silent moment, as he holds the jar underneath the tap of the settling tank and turns, hoping that something comes out (and if it does, that not *too* much comes out at one go or else there'll be another mess and more hell to pay). He did put the plastic sheet on the floor, didn't he . . .?

And when he comes to the end, and his jars, amply filled, are displayed in serried ranks on the surfaces around the kitchen, proud testament to the labours of both bees and keeper, he allows himself a moment of congratulation, of pride even, of hope that he might not be such a Bad Beekeeper after all. Until he hears the question:

'You did clean the tap on that tank where the dog licked it, didn't you?'

Author's note: I should add here that the scenarios described above are purely hypothetical, and any resemblance to people or events in real kitchens is purely coincidental.

And yes, I did.

Chapter Five

THE THREE D'S

*Disease, disaster and death: the calamities facing
the world of bees, and (some say) a world without them. The
nasty little* varroa destructor. *And why woodpeckers sound
like they're laughing.*

By now you may be thinking that there's really not an awful lot to this beekeeping malarkey. So long as you take care not to make the stupid and easily avoidable mistakes that Bad Beekeepers tend to, then things – and the bees – will pretty much take care of themselves.

Well, no. In years gone by you could pretty much put a hive at the bottom of the garden, leave it to its own devices and take the honey off once a year. But these days I reckon that the primary purpose of beekeeping is not the production of honey, but simply keeping the little darlings alive.

That's because the world has become a more complicated place; not just for us but for the bees as well. Disease has always taken its toll on bees, just as it has with everything else. But now the diseases, or rather the pests, have become cleverer.

One of the first things people say to me about beekeeping is, Oh, the bees have got a few problems, haven't they? Aren't they all dying out? Well . . . yes, but not really.

They have had a big problem in America with what's now called Colony Collapse Disorder, or CCD, when thousands upon thousands of colonies not only upped and died, but also completely vanished. Commercial beekeepers would turn up at their mobile apiaries and find the beekeeping equivalent of the *Marie Celeste* on a massive scale. Scientists in the United States have been puzzling over this one for a couple of years now, and may have it down to a specific type of virus which has found a weakness in the bees' makeup. The roots of the disease are natural. But here's the thing: the cause of the disaster that threatens the whole commercial beekeeping industry in the US is *man-made*.

The clue is in the word 'commercial'. The big money in American beekeeping comes not from honey, but from pollination. The growers of almonds in California, or blueberries in Maine, or apples in Washington State, all need bees for their crops to survive. They're the ones who pollinate the crops and thus make sure that the fruit develops.

It's simple. No bees, no pollination. No pollination, no

crop. So fruit growers pay huge fees to commercial bee-keepers for the services of their colonies. And as a result, the bees are transported vast distances, thousands of miles, from one crop to another.

All of which wouldn't be so bad but for two things. No one likes a diet consisting of just one source of food. How would you like to eat oatmeal, and only oatmeal, for three months, before switching to white bread, and only white bread, for the next quarter? Add to this the stress of being shipped long distances in lorries to different parts of the country. And the fact that they're jammed and slammed hither and thither, for commercial beekeeping doesn't allow any time for TLC. The hives have to be moved and resettled to deadline. And to make sure that they're ready on time, the bees' breeding cycle can be artificially stimulated so that there are enough of them set to fly – and pollinate the orchards – in early spring.

For the bees, all of this can make for a pretty miserable life. Stress can make anyone more susceptible to disease. And that's part of what seems to have caused CCD.

The added mystery, though, is in the manner of their passing. In many cases they simply seem to leave *en masse*. Some bee farmers in the United States and Europe argue that the use of neonicotinoids and other pesticides has damaged the bees' ability to navigate, so that once they leave the hive to do what comes naturally, they can't find their way home.

Alternatively, there may be more of a philosophical

explanation. If you're a bee living even a short life in such conditions, you might get to thinking, What's the point? Now, of course bees don't think as such. But they do have a collective intelligence, a way of making decisions on a communal basis. And it may just be that they calculate that life in the hive, once a virus begins to take hold, is untenable. And therefore it would be better for the colony as a whole to up sticks and seek shelter elsewhere. Whatever the root cause, it has clearly been bad beekeeping on a monumental and tragic scale.

At the same time that CCD was wreaking havoc in the United States, beekeepers in Europe were experiencing major problems as well.

French beekeepers had claimed that, over several years, a third of their stocks were wiped out by the use of imidacloprid, a pesticide used on sunflowers, and those colonies which survived produced less than half of their usual crop of honey. No one ever managed to prove that imidacloprid was to blame, but the French banned it anyway. There have been drastic losses since the turn of the century in other countries: Sweden, Greece, Italy and the Czech Republic.

And then in 2008, catastrophe. According to the world beekeeping association Apimondia, thirty per cent of the entire stock of Europe's bee colonies died. Four million hives. Fifty per cent losses in Slovenia. In southwest Germany, four out of five colonies perished. At this rate, Apimondia's president reckoned European beekeepers could only stay in business for another eight to ten years.

British beekeepers also had a terrible winter in 2007–8. Almost a third of the UK's hives failed to survive through to spring, and many beekeepers' stocks were wiped out.

All this raised fears that Colony Collapse Disorder had struck on our side of the Atlantic. But since we don't have the same pollination industry here, that clearly couldn't be the case. The cause, in Britain at least, may have had a lot to do with the weather. We had a particularly wet winter that year, the second in a row. And that did for many colonies. Bees don't mind the cold. They can withstand external temperatures well below zero. But what they really can't deal with is damp. So with two wet winters on the trot, a lot of hives bit the dust.

Normally a beekeeper would always expect to lose a small proportion of their colonies. It just happens. As the weather gets wintry, the bees reduce in number to about ten thousand, and form a cluster around the queen to maintain a constant temperature of about thirty-three degrees Celsius, whatever the conditions are like outside. Even if they have plenty of honey stored up to keep them going, strangely they may not find it. Bees tend to move up vertically in the search for supplies, and often just miss the life-saving frames of food half an inch away from them. It's quite odd, and of course rather sad, finding them this way. You open up the hive in early spring and find the bees static, as if caught in a freeze-frame.

THE BAD BEEKEEPERS CLUB

The good beekeeper does his or her best to keep the bees alive through winter by feeding them sugar syrup in the late summer, and then topping them up with something soft and sweet like baker's fondant (the stuff they use to make icing) in January.

The Bad Beekeeper wants to do these things but kind of forgets, or more likely leaves it too late, or sides with the purists who argue that feeding bees isn't really correct and they should be strong enough to survive on their own. Ironically, even the best-fed bees can up and die on you, but it makes sense that a better-fed bee is more likely to make it through to the spring.

Anyway, when the disastrous winter struck, people wondered if our bees had had it altogether, and out came the famous quote attributed to Albert Einstein:

> *If the bee disappeared off the surface of the globe then man would only have four years of life left. No more bees, no more pollination, no more plants, no more animals, no more man.*

This is nonsense, for two reasons.

First, there's no hard evidence to suggest that Einstein ever said anything of the sort. There is no written record of such a statement. The earliest source that can be found is from a group of French beekeepers in 1992, who might just have had secret and unique access to some hitherto unknown paper written by the great man, but who, more likely, just made it up.

Second, it doesn't stand up. True, bees are responsible for the pollination of a great deal of what we eat, particularly fruit. But other insects can do the job as well, and other big crops like wheat, corn and rice are pollinated by the wind. There were no honeybees at all in North America until European settlers took them there. So how did anything survive until then? True, life would be significantly less pleasant without bees. But we'd still survive.

Anyway, one year after the terrible winter of 2007, conditions were even harsher. It was much colder in many parts of the UK. And guess what? More bees survived. In 2008 the average national losses were just under a fifth; although the figures for northern England were still higher, at almost one hive in three. Still too high, but not as catastrophic as before.

It's not just starvation that'll get your bees in the winter. There are any number of pitfalls awaiting beekeepers good and bad. Most notorious is the varroa mite. Its formal name is *varroa destructor* – a perfect description. It's a little red flea-like insect which hops on the back of the bee and feeds on its blood. It grows in the larvae of the brood, and so has the perfect place to live and breed. It can multiply one-thousand-fold in a single year. It's bad enough on its own, but it also carries no fewer than fifteen viruses, some with villainous-sounding names: acute paralysis, black queen, slow paralysis and Kashmir. Nasty, aren't they?

The one we see more of here in Britain is the deformed wing virus, which irritates bees so much that they chew their own wings. Even when the varroa mite has been cleared out of the hive, the virus can stay behind, and you will occasionally find deformed bees with virtually no wings left moping around the frames.

There was quite a good treatment for varroa until a few years ago. But the mite became resistant to it, and even stronger than it was before. So now beekeepers have had to learn to live with the invader by controlling it rather than getting rid of it altogether. They usually do this by administering a thymol paste, which smells so bad that the mite can't stand it and falls to the bottom of the hive. The good beekeeper uses wire mesh for the floor of his hives, so that the varroa fall through it and to their doom. The Bad Beekeeper has stuck with wooden floors, so that the mite can simply get up and climb back into the hive to resume its dastardly work.

Although we haven't got it here, there may be a connection between Colony Collapse Disorder and our own problems in the UK. For among the potential culprits behind CCD is what's called the Israeli acute paralysis virus, which can be carried by that one-stop shop for all sorts of unpleasant disease, the varroa mite. Its particular trick is to afflict the bees with shivering wings, and then paralyse them. Eventually – and here's a crucial clue – they die *outside the hive*. With tens of millions of bees from all over the United States working and mixing in the almond orchards of

California, it may be that they carried the virus back with them to other destinations – and other bees – right across the continent. With their immune system depleted by varroa anyway, and in their stressed condition, they would have made easy victims.

But that's just one theory. Other scientists pin the blame on a new form of an old enemy: nosema. It's a parasite which affects the bee's intestines, rather like dysentery for humans. Usually it can weaken a hive and reduce its honey production, but it's fairly easily treated. Now though, new and improved from China, comes *nosema ceranae*, a much more vigorous form of the disease which scientists in Spain believe may be responsible for many of the heavy losses on our side of the Atlantic.

As the man from *The X-Files* said, 'the truth is out there'. It's just taking time to find it. But you get the picture. And it's rather bleak.

As it happens, Mulder and Scully had a bit of a thing about bees. In the film *X-Files: Fight the Future*, the two agents are set upon by a massive swarm of bees carrying 'the Black Oil' within their venom, which was derived from genetically altered corn crops. And in earlier episodes of the TV series, our heroes had several encounters with bees carrying the smallpox virus. You'd have thought they could have come up with something more credible than that – like a conspiracy to wipe out the world's entire population of bees so that aliens could take over. Hang on a second . . .

In real life though, there's a whole host of other threats waiting to strike the unsuspecting: chalkbrood, sac brood, stone brood, neglected drone brood, bald brood, chilled brood, acarine, and the deadliest of all – American foulbrood. This last one is the worst. It is the Black Death, the Ebola virus, of the beekeeping world. It attacks the unborn bees and, if not treated in time, will kill the whole colony. It is highly contagious, and once it's landed in the woodwork of the hive you can't wash or scrub it away, or even freeze it to death. Its spores can survive for up to forty years. So the only thing to do is destroy the colony. Kill the bees, and burn the hive. There is no other way. Fortunately, it doesn't happen very often. But it's one that strikes a chill in the hearts of beekeepers everywhere, good and bad.

Having said all that, it's not all doom and gloom. We've had heavy losses before and survived. One hundred years ago, Isle of Wight disease (because that's where it came from) spread unchecked and wiped out pretty much the entire bee population of Great Britain. Stocks recovered thanks to imports from abroad. With everyone else losing their bees, it might not be so easy to replenish them today. But in spite of all the problems, there are still nearly a quarter of a million hives in the UK. We just have to keep an eye on them.

They do say though, that the bees' greatest enemy is none other than the beekeepers themselves. The bees would be doing perfectly well on their own in the wild, if they had been left alone. But there are, it's feared, no feral bees in

Britain any more. They've all been wiped out by varroa, which we've helped to spread. And too many of us tend to go tearing into hives with our large clunky leather gloves, squashing and maiming bees in the process. To be fair, you probably need to wear leather gloves when you start, to protect your fingers. But in terms of sensitivity, it's a bit like trying to eat your breakfast wearing boxing gloves.

And in my humble opinion, we don't help ourselves here in Britain by messing about with the gene pool. We're forever trying to improve the temperament and productivity of the hives by introducing new queens. You can get them from all sorts of different places: all over the EU, especially Cyprus; Australia; New Zealand; Argentina; even Hawaii. It's not complicated. You order them over the internet and they turn up in the post. Just a small aerated container about the size of a cigarette packet, with a queen inside and half a dozen workers to attend to her on the journey. When they arrive, the workers are done away with and the queen is introduced to the colony.

This is all well and good, and the queen may prosper. But often they don't. I once had a queen from New Zealand, who produced bees with a good docile nature, but she didn't survive her first winter. And some would argue that we're not exactly strengthening the gene pool here. It's a bit like taking a farmer from southern Italy and asking him to work in the Yorkshire Dales. What we should be doing (and here I doff my cap to the geneticist and veteran beekeeper Dorian

Pritchard, who explained it all to me) is breeding our own local bee, one that will survive and thrive in its own particular environment. I wouldn't expect the bees that do well for me in Buckinghamshire to prosper in the Highlands of Scotland. So why do we think that bees from a hot Mediterranean background will survive in Skegness?

Dorian's idea to help protect the future of bees in Britain is for us all just to select the best variety from our own bees. In other words, if you have a queen that has survived three winters, she is clearly hardy enough for your local conditions. She is, in effect, now your local bee. If you breed from her, your colonies stand a much better chance of making it in years to come.

The problem is that beekeepers can be impatient. If your queen snuffs it in the spring, you need to replace her as soon as possible in order to keep the workforce growing (2,500 eggs a day, remember?). Otherwise you'll never get the numbers you need to produce a reasonable amount of honey. And if you've only got one or two colonies, the only answer – it might seem – is to get on the phone and order one in from wherever you can get it. This might get you out of a hole, but it will also help to perpetuate any genetic problems.

And if the humans don't get to the bees, there are plenty of others who'd like to have a go. Mice, for instance. They love to move into the hive for the winter. It's nice and warm inside. They're safe from predators there. They can nest on the ground floor while the colony clusters up above them, and if they get peckish they can eat their way through the wax and the stored honey during the colder months. When you open up the hive in the spring, you'll find that they've made big fat holes in all the lower frames, and the whole place smells of mouse pee.

Woodpeckers love to drill their way through the walls of the hive and suck out the larvae. You can come back to a hive after the winter and find a perfectly round three-inch hole in the woodwork, and a dead colony inside. Beekeepers will tell you that the bird's cackling song is Woody having a good laugh at their expense. Badgers have been known to come along and knock the whole hive over to get to what's inside. Wax moths love to lay their eggs in the frames. And in the summer, the wasps and hornets will do their best to get in and rob the place out. Hornets can be particularly devastating. The Japanese variety, two inches long, can decapitate forty bees in a single minute. When word gets around, a raiding party will invade a hive and lay it to waste, killing all the inhabitants in a couple of hours. It's the larvae they're after.

And guess what? Some of their cousins are over here. Asian hornets, probably from China, have been spreading in south-western France since 2004, and they're heading our way. Eleven hundred nests have been discovered so far, some as far north as Brittany. It's only a matter of time . . .

Coping with all the maladies that can afflict our little *apis mellifera* can take a lot of work. Someone asked me a while ago if they could simply buy a hive of bees and just put it in their garden. Their motives were noble: they wanted to do their bit for the environment and support the bee population. But I had to put them off. These days, bees can't be left to fend for themselves. Sooner or later, they'll succumb to disease, and infect other people's hives into the bargain. And before they expire, they'll probably upset the neighbours by swarming a couple of times as well.

So, even Bad Beekeepers need to practise their own three D's: dedication, discipline and determination. One because it's not a hobby to be undertaken lightly. One to guarantee that the things that need to be done, get done. And one to make sure that when things are going wrong, as they surely will, you keep yourself going.

Did someone say beekeeping was relaxing?

Chapter Six

SWARM!

A swarm of bees in May is worth a load of hay.
A swarm of bees in June is worth a silver spoon.
A swarm of bees in July is not worth a fly.

Bees, eh? When they're not dying on you, they're trying to get away. Talk about ingratitude. You feed them, you house them, you nurse them, and what do they do when they get the chance? They leave you. In a swarm. It's the most widely misunderstood element of beekeeping. People find it very hard to understand. So let's try to make it simple. You ask the questions.

What is a swarm?

A swarm is a large number of bees, which emerges in one vast cloud from a beehive, in search of a new home.

Why are they doing that?

Because they need more space. During the spring and early summer, the queen lays a lot of eggs. When they hatch out, things can get a bit crowded. So the decision is made for the queen to leave with half of the colony, more or less. The worker bees start giving a number of eggs special attention, so that a new queen will be born. When these are mature enough to be sure to survive, the old queen leaves with thousands and thousands of bees, and they go off and look for a new home.

So that's what's happening when they swarm. But it's also the way that the bees, as a species, can reproduce on a broader scale. Because if they didn't swarm, there'd be no way they could increase their numbers. The number of colonies would remain the same, and eventually the species would die out. And we all know what would happen then, don't we, Mr Einstein?

Isn't it frightening?

Not really. Once you understand what's going on, it can be the most wondrous part of beekeeping. Because they have no home or territory to protect, the bees are not at all aggressive. They're just sort of lost. You could probably stand in the middle of a swarm and not get stung, unless one of them accidentally bumps into you and thinks its life is threatened. Though I should add, I haven't tried this myself yet.

Where does the swarm go?

Usually, and with a bit of luck, it ends up hanging from the branch of a nearby tree. I have known them to go into bird boxes. Some took up residence right at the top of our local church. A friend found a swarm between the window pane and the shutters of a villa in Italy. And they do have a bit of a thing for chimneys, of which more later. I once got a call-out for a swarm which had gone to the races and settled under a chair in the Royal Enclosure at Ascot. Sadly by the time I got my gear together, they'd disappeared. Maybe they didn't like the hats.

For the beekeeper, swarms can be a good thing or a bad thing, depending on their provenance. Of course, the bees would not accept that they belong to anyone; though they clearly understand the concept of ownership, a point they are happy to prove every summer when you want to take their honey away.

But what I mean is, you don't want your own bees to swarm, even though it's the most natural thing for them to do. Because if they do go, that hive has lost half of its work-force, and depending on the time of year, you can pretty much kiss goodbye to your chances of getting a decent amount of honey from them. And there's the neighbour issue. For some strange reason, people tend not to like it when tens of thousands of bees swirl in a cloud over the garden when they're not expecting it, especially when they're having drinks with guests on their patio. And you rather feel like an apologetic parent chasing after a mischievous child in a supermarket, unable to catch up with them as they leave havoc in their wake.

A few years ago I followed a swarm from a hive on my own property, into my next-door neighbour's garden, and then into the following neighbour's garden. Then they went up the hill, through another property and over a swimming pool, where I had to reassure a – quite understandably – concerned parent, whose children were swimming at the time, that the bees were just paying a flying visit and would be moving on soon. Back down through my own garden again, across the lane to the big house where the owners were hurriedly shutting all the windows, deaf to my shouts of, 'It's okay, it's just a swarm . . . they're my bees, sorry, sorry!', and eventually out through their meadow, over a hedge into a large overgrown field. There I had to stop and wave them a fond farewell, as they wove their way towards the next village. I never saw or heard of them again.

On the other hand, there's nought so joyful, in my book at least, as collecting a swarm of bees that has arrived out of the blue, and doesn't already belong to you.

Ownership of bees, as we humans understand it, is fairly simply sorted out. It's generally accepted that they belong to the person or persons on whose land they have settled. If they've taken up residence in your garden, they're your bees. You might not want them there, but they are yours. So if my prize bees up and leave and fly over the fence into your property, I do not have the right to take them back. Unless you let me. And be honest, you don't really want them, do you?

So, going to pick up a swarm is always a joy. Though I do spare a thought for the poor unknown beekeeper who has lost their bees. Aah. There, there.

On the other hand, the principle of swings and roundabouts seems to apply fairly evenly in my experience. The excitement of picking up a new colony compensates for the aching feeling of disappointment when you see your own bees streaming away. Or worse, when you get the call from home that the bees have swarmed AGAIN and you're a hundred miles away. More than once I've had to issue instructions over the phone to family and willing neighbours about how to get the blighters back.

'Okay, where are the bees now? Hanging from an apple tree? Good. Go and get the stepladder. The one in the garage. Then go to the shed and get the skep out. A skep? It's the thing that looks like a straw bucket without a handle. You can't miss it. And don't forget to put the beesuit on.

'Now, hold the skep underneath the part of the tree where the bees are clustering. Grab the branch that they're hanging from and, when you're absolutely ready, give it a sharp tug. The bees should all fall into the skep. It might be a little heavier than you think, though, so be prepared. Then place the skep with the bees inside it *upside down* on the sheet on the ground. Which sheet? The old white one in the shed. The one I forgot to tell you about. And when you've done all that, call me back.'

Half an hour passes while you sit in your hotel room. Waiting, hoping, wondering.

The phone rings.

'What? Yes? Good. And you got them on the ground. Great. That's terrif— what do you mean, they've all gone back up in the tree?'

You can try to get round the swarming problem using various tactics. Some beekeepers like to clip the queen's wings so that she can't fly and leave the hive. This has inspired the same sort of debate in beekeeping as women priests have in the Church of England. It's also not recommended for novices or Bad Beekeepers, as not only could your queen end up without any wings, but one slip of the shears and you could end up without any queen at all.

There are a couple of other tricks you can play on a colony. Splitting it in two for a while can make the bees think they've already swarmed. Or you can perform a 'shook swarm', which is exactly what it sounds like. Again, you con the bees into thinking they've swarmed by shaking them into new premises. It's quite dramatic and involves lots of confused and possibly angry bees flying all over the place and is hence a favourite for the Bad Beekeeper.

I always leave a couple of empty hives in the garden in the hope that the bees will seek new accommodation close to home, and move in there. The trouble is, others find the hives quite comfortable as well, like mice or slugs. One colony of bees moved into what we call a 'bait' hive down by the shed, only to find a pair of nesting blue tits and their offspring as neighbours. I'm not sure who would have been more confused.

Perhaps the most exciting thing I've witnessed as a bee-keeper, though, was the day a swarm landed in an empty hive on the garage roof.

I happened, fortunately, to be near the front of the house when I heard the ominous roar that every beekeeper recognises instantly. Funnily enough, it can sound a bit like a train. I looked out of the window, and there they were. Tens of thousands of them pouring down in a torrent, as if sucked down by a giant invisible vacuum, in the space of a minute. It was, quite literally, a joy and a wonder to behold.

But sometimes, however hard you try, you lose them. The biggest, most beautiful and easiest swarm I ever caught was one of those that got away. Eight o'clock on a warm evening in June. Watching the telly and the phone rings. It's a friend in the village. There's something in his front garden, and it's buzzing, and would I come and have a look?

Like a Kansas storm-chaser, I am tooled up and ready to go in a matter of minutes. As the light begins to fade, I arrive to find it: a large, peaceful, dark brown cluster of bees hanging from a young silver birch at the perfect height. They've even got a picnic table just next to it, so I can stand on that and pull the branch down like a massive wooden lever to get at the bees. It's as if a large invisible net is being released beneath them. Down they fall *en masse*, almost like liquid, filling up the skep. Quickly inverted and left on the table, and the job is done in minutes. Textbook. And then . . .

I couldn't re-house them the next day. Something came up, and I didn't have the frames ready for a new hive. Never mind, you can keep bees in a skep for ages if you want to, they'll be all right there. For convenience's sake, I once left a skep full of bees inside another hive for a whole winter. Didn't mean to – I just sort of forgot.

So, full of confidence, mid-afternoon the following day, I went round to pick up the lovely swarm. There were a few bees flying around the entrance to the skep, but it looked a little quiet. Sadly, too quiet, for when I lifted up the basket to check inside, it was rather light. Too light. There was not a single bee left inside. They'd gone ten minutes earlier, and wafted off into the distance.

You remember that feeling you got when you were young and you met someone really, really nice and attractive and you got on really well and it was really exciting and you told everyone and you thought that this was going to be sooo great . . . and then they turned around and, out of nowhere, dumped you? That was how I felt. Spurned. Jilted. Bereft. Utterly depressed. I'd had a beautiful prize in my possession and I'd let it slip through my fingers. Even when I think about it today, I jus' can't help comin' over all country. Move over, Charlie Rich . . .

Hey, did you happen to see
The most beautiful swarm in the world?
And if you did, was it flying? Fly-ing . . .
Hey, if you happen to see
The most beautiful girls that waltzed out on me
Tell them I'm sorry
Tell them I need my bees back
Oh, won't you tell them that I love them

Moral of the story: in beekeeping as in life, timing can be everything. And the Bad Beekeeper usually waits too long.

The phone rings more in summertime. People call, total strangers sometimes, asking for advice. Usually I'm only too happy to give it. Though once word spreads that you know what to do with bees, it can be open season. The conversation goes something like this:

Hello, is that Mr Turnbull?
Yes.

I wonder if you can help me. I've got a swarm of bees in my garden and I don't know what to do.

OK. Are they definitely bees?

Oh, yes. I can see them flying in and out of the roof tiles.

That's unusual. You're sure they're honeybees? Not bumbles? Or wasps?

Oh no, they're definitely bees.

So, off I go, gathering all the requisite gear. Skep, sheet, smoker, syrup, suit, gloves and ladder, only to find that more often than not, they're not honeybees at all, but mislabelled bumbles. Once I got called out to a man's house ten miles away. He said he had a swarm on his patio. A few years back, he told me, he'd been Treasurer of my local Association, but only as a gentleman beekeeper – looking but not touching. So I had to go as a matter of honour, for the good of the Association and all that. There they were, flying merrily in and out of a nest behind his patio. Not bees, but wasps.

Now, some beekeepers will charge for a call-out to a swarm, to guard against just such an eventuality. It's usually only a tenner or so, to cover the petrol and a bit of their time. But I don't. I like to put people's minds at rest about whatever's flying about their garden. And I'm always curious to see how and where other people live. You'd be amazed how far inside someone's home they'll take you if you're wearing a veil and carrying a skep.

I certainly always have to go if I get a call in my own

village, because there's a pretty good chance that if there's a swarm, it'll be one of mine. Chimneys have been the in thing for bees recently. They just love getting in there where it's dry and dark, unaware of the panic they're creating in the house below them. Do they realise, I wonder, just how difficult it is to get them out of there?

I spent a whole Sunday morning trying to remove one lot from a chimney in the village, where they'd parked the night before. It was almost too late, as once the queen has started laying in her new home, they won't leave the brood unprotected, and won't budge. Timing nearly did for me again, but fortunately in this case, they hadn't had long enough to manufacture new wax cells. Once I got the call from the understandably perplexed householder, I enlisted the aid of my beekeeping mentor, the ever wise and willing Christopher. It was he, you'll remember, who saved me from the bee inside my veil on that first fateful day at the apiary. Over the years he's been a constant source of useful advice and encouragement and as ever, when I didn't know what to do, I turned to him for help. It wasn't long before we were both getting down to the job, quite literally. For the only access to this particular chimney was through a vent in one of the bedrooms. There, we did what beekeepers do best: we puffed smoke. Resolutely, up the chimney, for the best part of an hour. We took it in turns. One of us would lie underneath the desk in the teenage occupant's room and blow and cough, while the other hung out of the window looking for any signs that the bees might be about to depart. They certainly took their time. Eventually, they did emerge, only to perch defiantly on a large cedar tree right next to the house, clear their lungs, and then try to get back in again.

Round Two took place in the garden. Out came a high-pressure hose, tied to a broom handle, administered precariously from a garden chair – just high enough to dowse individual members of the swarm as they tried to work their way in through a different route, this time under the tiles. Like humans, chickens, dogs and pretty much everyone else, bees don't like being sprayed with water. On the other hand they do like chimneys, and this lot were determined to get back in there, come hell or hose water. The standoff continued for an hour or more. We sprayed, they left. We stopped, they came back. It was truly a battle of wills. And time, Christopher reckoned, for the nuclear option.

Sulphur strips are normally used to keep empty hives clear of the wax moth in the winter. The moth itself isn't a problem. But it lays eggs in the beeswax comb, which become larvae, which tunnel through the wax and turn it into a useless mess. Sulphur strips do away with the moths, and a whole lot of other things too. You pile your empty hives together, light the strips underneath, and withdraw quickly. Anything hanging around above has pretty much had it. It's not a pleasant thing to do, and not a pleasant substance to breathe in either. It smells horrible, and it's poisonous. But sometimes it can be the only remedy. In the old days of skep beekeeping, when colonies were housed in simple straw buckets, sulphur was what the peasants used to kill the bees before removing their honey. We weren't planning to bump this lot off, just to send a message: Go somewhere else.

So we lit a cluster of three strips, shoved them in the chimney vent, and beat a quick retreat from the fifteen-year-old's bedroom. In many teenage boudoirs, you might be hard pressed to tell the difference between sulphur fumes and the normal adolescent aroma. But on this occasion the only noxious fumes were those emanating from the sulphur, and at last the bees decided they didn't want to come back. For a while, they hung defiantly in the top of the tree, as if to taunt me. A few days later, they disappeared, never to be seen again.

The greatest recovery challenge so far, though, was down my friend Jeremy's chimney. The bees had been there for a while, for years in fact, when he told me about them. Their arrival had conveniently coincided with that of his mother-in-law, on a visit from Poland. As she unpacked and settled in, they did too, hurtling down the chimney and into her bedroom. The window turned black as they tried to find a way out. Perhaps they were intimidated by the sight of a Polish mother-in-law, but at any rate eventually they withdrew to the chimney. Not knowing what to do, and deciding diplomatically that mother-in-law should take precedence over a swarm of strange bees, Jeremy sealed the chimney at top and bottom.

Not that that was enough to stop the invaders. As only they can, they found a small hole in the mortar of the chimney stack to get in and out of, and settled happily into their new abode. Years passed, and the bees could still be seen flying to and fro. They might be there now but for the fact that Jeremy and his wife realised that one day they might want to sell the house, and that a working chimney or two might be an additional asset, while a colony of bees as sitting tenants might detract slightly from the overall value.

For some reason, they approached me with their predicament. Not because they thought I was any good as a beekeeper, but because I was the only one they knew. Perhaps they were hoping I might put them in touch with someone who knew what they were doing. But this was an opportunity too rich with potential for disaster for any self-respecting Bad Beekeeper to pass up.

So, I mulled and pondered and puzzled as to how to get bees out of a chimney where they had been resident for several years. For there would be not only bees, but brood and honey and lots of wax, each adding its own complication. For even if you managed to remove the live bees, the brood – the eggs, larvae and pupae of the unborn – would eventually die and decompose and possibly smell rather unpleasant, while the honey might begin to ooze out of the comb and seep through the walls. And the wax? Think chimney. Think fire.

Think flames roaring up and down as the wax melts and combusts. Think fire brigade. Think insurance claim.

Actually, let's just stop thinking about it.

And anyway, how could you lure bees downwards, out of a chimney, when in all likelihood they would be high up, and we know that they always fly upwards towards the sun when leaving the hive. You can't tempt them down with food. And smoking them out wouldn't work either, as we know from our previous experience that they would only try to go straight back. So I mulled and pondered and puzzled a bit more, and then did what all great minds do in such a situation. I looked it up on the internet.

Eventually I found a website in Australia where someone explained how he had done the very same thing, and the solution was elementary – you just had to look at it from a different angle. Instead of forcing them down, you simply let them fly out of the top, like they had been for years, and then (here's the key bit) *stop them getting back in.* Of course! Why didn't we think of that?

Actually the answer's not quite that simple, as you have to give the bees somewhere to go when they fly out. And that means putting a beehive on top of the chimney. With some bees already in it.

It would have been hard enough if the chimney had been relatively normal, at the apex of the roof for example. But no, this one had to be somewhere else. In your mind's eye, go to the top of the house. Sit comfortably on the ridge if you can.

Then look down to your left, almost at the edge of the building, with a thirty-foot drop beyond it. That's Jeremy's chimney.

When I first arrived to take a look, I nearly got straight back in the car. I'm not brave and I don't do heights. Moreover, I couldn't see how anyone brave who did do heights could manage it. Jeremy, however, has been about a bit. He's big and tall and daring. He rode motorbikes in his youth and now crews on yachts, so he's used to risking his life and doing things at odd angles. And he'd been up there to put the slabs on the chimney in his albeit unsuccessful attempt to seal it. And he'd done it with bees flying around him. So clearly it could be achieved, and here was the soul courageous enough to take on the challenge. All I had to do was stand behind him and give instruction. What could possibly go wrong?

I had some bees to spare, a small swarm out of one of my own colonies. The hive on the chimney pot would have to contain some, in order to show the evictees that here was a safe, habitable spot to move into. All we had to do was get them up there.

It was rather like a mountaineering expedition. First we climbed a ladder up to the gentle slope of the extension at the back of the house. Then a more direct ascent to a flat-roofed section, before the final push to the top. Not too challenging, unless you're wearing a jacket and veil and carrying a brood box with bees inside it. As I knew that the really difficult bit was not going to fall to me, I took it upon myself to be Sherpa Tenzing to Jeremy's Hillary, and insisted on carrying the bees. They were mine, after all. For his protection, I'd given Jeremy one of my own beekeeping suits; one I'd run a marathon in, with BILL stitched in very large letters on the front. With a bit of luck I reckoned the bees might be suitably duped and, in a case of mistaken identity, sting him rather than me.

It was a still, sunny summer's evening as we reached the top and stood for a moment on the apex, overlooking the main street of the Oxfordshire village. Down below, a couple of wags loitering outside the pub cried, 'Don't do it!' We must have been an odd sight.

Here self-preservation overtook valour on my part; never a close race, to be honest. I let Jeremy shuffle on ahead of me and passed the box to him, before squatting back down and clinging to the ridge, like a small child riding a beach donkey for the first time. I'd armed him with a lit smoker and a syrup spray as he approached the chimney stack. First he had to glue in the trap – a cone of firm mesh wire which would cover the bees' exit from the chimney. A bee-sized hole would let them pass one at a time out of the pointed end, but they would not be able to get back in again. Devilishly cunning.

Once he'd fixed the trap, Jeremy then had to shuffle down the roof again; this time on his backside, carrying the hive, while an increasingly confused reception committee of chimney bees gathered around him, wondering why all of a sudden they couldn't get home. Could this giant invader have anything to do with it?

In a matter of minutes the job was done. Hive screwed to platform, fixed to chimneystack; roof secured with wire in case of high wind; cone in place and, it seemed, working.

Now observing such a scene, you might conclude that the successful performance of such a task could hardly be the work of a Bad Beekeeper, for it required some skill and courage and a sense of endeavour. And you'd be right. Which was why I had little to do with it. While Jeremy had been risking life and limb, I had been at a safe distance, kneeling astride the roof, telling him what to do. A bit like a General, really. Mission accomplished, we climbed back down the mountain and left the bees to it.

The next phase of Operation Clearout could not start for several weeks. We had to wait for most of the bees to leave. Then it would dawn on the queen that her colony was dwindling, and that she was being deserted. She would then stop laying eggs, her remaining brood would hatch out, and the last bees would leave home. That would take at least three weeks. Once the chimney was clear, we would remove the trap and do something you might not expect: we'd let the bees back in again. By now the queen would have left, or died, and there would be no reason for them to take up residence again. What they'd do, we hoped, was take away all the honey stored in the chimney. This would then leave just the wax, which would eventually be eaten up by moths. No, really.

But would it all work? Well, time passed, and the world turned, and for a few weeks I turned with it, literally circumventing the globe for my part in a new version of *Around the World in Eighty Days*. The BBC was remaking Michael Palin's epic journey, filmed twenty years previously, for Children in Need, but this time with six teams over six one-hour programmes. My mission was to fly to Mongolia with fellow newsreader Louise Minchin, pick up a carpet bag from the previous team (Julia Bradbury and Matt Baker), and take it with us through Russia, down to South Korea and across the Pacific to California. We were away for more than three weeks, and carried the bag for ten thousand miles, before passing it on to the next pair of travellers, John Barrowman and Myleene Klass.

During this time I'm sad to say I did not encounter a single bee. Well, you won't find many in the middle of an ocean, to be honest, though I did pick up a jar of local honey at one station stop on the Trans-Siberian Express.

The bees in the chimney, though, were never far from my mind. Every day I wondered whether our cunning ruse was succeeding. As we rattled across the vast empty plains of northern Mongolia, I wondered, Where were they now? When I lay awake with a rumbling stomach in a Siberian guesthouse, I asked myself: Had they emerged? I took my time contemplating this question, as the surroundings at that point were particularly awkward. The house we were sleeping in – or trying to sleep in – had no running water, and hence no indoor loo. The dunny, a wooden hut housing a deep hole in the ground, was outside. Outside past the chained guard dog, at the far end of a dusty farmyard carpeted with freshly laid cow dung. And it was dark. And I had no torch. So, listening to my digestive system doing battle with the freshly slaughtered but rather oddly cooked (and hence almost inedible) lamb that we'd eaten for dinner, I tried to distract myself by thinking of the bees emerging from the darkness of the chimney. Would it affect the taste of their honey, I wondered. How were they settling into their new home?

As we were out of contact with our own home for most of our time away, there was no way of telling. And on the rare occasions that we were able to phone, it wouldn't have done to interrupt the flow of sweet nothings by asking for a progress report on the flight of the chimney bees.

'Missing you too, darling, more than words can say. Any news from the rooftop?'

When, finally, we had done our bit for Jules Verne and flew home from Los Angeles, one of the first calls I made was to Jeremy for a progress report. He got out his binoculars. Yes, the bees had emerged from the chimney. Yes, they seemed to be using the hive on the stack. But for some reason, there was still traffic flying in and out of the conical wire trap we'd glued to the hole at the top of the chimney. Lots of bees stuck inside the cone trying to get out. This didn't sound good. But . . . hang on, they're not bees. They're wasps.

Wasps! The nemesis. The cursed arch-enemy of the honey-bee. Damn their yellow and black stripes, they were going to ruin the whole project. To hell with unpacking, I had to get up there. Within the hour I was up on the roof for a closer look. Having spent ten days at sea, I now felt confident enough to stand up straight on the apex, rather than kneeling and shuffling pathetically along. And there they were, the black and yellow devils, clustered together in the cone and, fortunately for us, unable to get out. If the wasps were going in, it meant the bees must all have vacated the premises, so we could now remove the trap with the pests in it. Some carefully administered fly-spray put paid to them, and the bees were free to go back in and rob out any honey remaining down below.

I should have taken them away a fortnight later, but hey, I'm not a Bad Beekeeper for nothing. Two whole months passed. As usual, I was distracted by other things like work and the beginning of the new football season (not necessarily in that order). It was only as autumn was approaching that I realised something really had to be done, or else the poor things would be spending the winter up there. And on the rooftops it can get rather breezy.

So on a mild Sunday evening, twelve weeks after we'd first taken the beehive up, Jeremy and I made the ascent to the rooftop for the last time. This was going to be a bit tricky, not least because his spare wooden ladder, which we'd used earlier for stage two of the climb, had broken. So the procedure went something like this. On the ground, suit up, veils on. Climb ladder to roof. Carry up lit smoker, sugar spray, equipment bag. Lift ladder up for transfer to stage two. Climb to upper part of roof. Pause while expedition photographer discovers batteries on camera have gone flat, and has to descend for fresh supplies. Move ladder back down to stage one. Replace ladder for descent to ground. Then do it all again to get back up to roof.

At last we were ready for the big lift. I was a little apprehensive, as the bees we were removing would be significantly heavier than the ones we'd put there in the summer. There would be more of them, and there would – we hoped – be all the honey they'd extracted from the chimney. Still, there was nothing for it now but to give it a go. I stuffed a strip of foam into the entrance to stop the inhabitants escaping, and we undid the screws holding the hive in place. A moment later, Jeremy had lifted the box away and had it in his arms, gasping with the weight. Think of holding a large cardboard box. Now fill it with fifteen large bags of sugar. Now carry it along the apex of a tiled roof, and down two ladders, taking care to keep it level.

It's no small achievement, is it? And that's what I thought as I watched Jeremy do most of it. Well, he's bigger than me. I just made sure I was in the frame at the right time, as the pictures were being taken.

Still, as we drove back down the motorway with our box buzzing with irritation in the back, I couldn't help but feel some satisfaction. Somehow, we'd managed to remove thousands of bees from deep inside a chimney some thirty feet off the ground, cleared out the contents and dispatched a crowd of wasps into the bargain. And we didn't get stung once. Maybe I wasn't quite so bad at this beekeeping malarkey after all.

Moving hives is one of the most fun and challenging things you can do in beekeeping. It requires planning, forethought, preparation and skill. Hence, for the Bad Beekeeper, plenty of opportunity to get it wrong.

Why, you might wonder, would anyone want to move bees anyway? Well, commercial beekeepers do it all the time, as they move thousands of colonies from crop to crop. That's how they earn their money, not through honey but through pollination.

I move my bees when I want to unite two colonies which may not be strong enough to survive on their own. Or if a hive that has been in the garden starts getting testy, I'll take it to the farm nearby, where I keep several other hives. There, out of sight and out of mind of the rest of the world, it's less likely to become a nuisance.

There are, of course, complications. The first is the three foot, three mile rule. Or the 0.92307692 metre, 4.8 kilometre rule if you're reading this in an EU country outside the UK. You can move a hive fewer than three feet or more than three miles, but not in between. Otherwise the flying bees, those who go out foraging in the last two weeks of their life, will get lost. Fewer than three feet, and they'll wonder why the front door has moved, but they'll still find it. Fewer than three miles and there's every chance that as soon as they're released they'll head straight back to where the hive was originally, only to find there's nothing there. Bees have a remarkable memory and sense of navigation. And without the flying bees, there's no fresh food supply coming in, which can put the colony at risk.

Once you've established a new site, you must prepare the bees for the journey. That means taking the roof and cover board off the hive, and putting a mesh screen on top for ventilation. Bees get stressed on a journey and their temperature can rise considerably.

Then you have to secure the hive by tying straps around it, from side to side, and, to be doubly sure, front to back as well. If you've got a super or two on top of the brood box, you may need to hammer in some serious-sized two-inch staples as well. While you're doing this, the bees inside will notice that something's going on, and depending on their temperament, may not take to it too kindly. It all adds to the excitement.

Then you reduce the size of the entrance with a thick strip of foam, so that just a few bees can go in and out at a time, and retreat to a safe distance for them all to calm down. A couple of hours later, you can close off the entrance and carry them away.

Dusk or dawn is the best time to move bees, when there'll be fewer of them flying. I always like to do it in the evenings, as it's warmer and there's more time to prepare. It's always rather exciting to think that you're removing fifty thousand bees in one go, in the back of your car. You can buy a sign that says CAUTION BEES IN TRANSIT, stick it on the back window and watch other vehicles give you a wide berth. And there's nothing like finding a bee on your thigh to concentrate the mind as you drive up the motorway.

My favourite move was the one that so nearly ended in disaster. It involved a hive that had been abandoned in a field, and adopted by me. As there weren't many flowers around, it wasn't a good site for foraging. I had to drive across a field to get to it, which wasn't terribly convenient for me or the farmer.

The slight complication was the size of the hive. It consisted of a brood box and two supers, and was going to be too big for me to lift into the car, and possibly too tall to fit in the vehicle anyway. What to do?

It is the genius of the Bad Beekeeper that, faced with two options, he will unerringly choose the wrong one. Not only that, but it will be the decision with the worst possible consequences. I've done it dozens of times, and it never fails. It's as if there is a magnetic fault in the brain which kicks in whenever the question-answering current is switched on.

Imagine you get in your car, and come to a junction . . .

Question to brain: Which side of the road should I be driving on?

Slight crackling sound due to surge of idiot current.

Answer: On the right.

So in this case, faced with a hive that was too big to move, I came up with the following options:

a) To abandon the exercise for the day, seek help and come back another time.

Or,

b) To split the hive into more easily movable sections, and carry on.

On paper, you might think that (b) looks like an inspired move. Of course it does. It's the fool's choice. And naturally, it was the one I took.

The thing is, beehives are not like solid objects that you can parcel up at will. They are in effect a living organism, not just a box containing tens of thousands of flying insects. So taking one apart would be like cutting a town in half.

Still, that's what I did. It was an ugly manoeuvre involving rather a lot of last-minute bodgery – trying to stick bits of mesh netting on in place of the extra travel screen that I hadn't brought. And at the crucial moment, I realised I was also short of a floor. Remember the bit about sealing the hive for transit? Well, one half was sealed only at the top, which meant that the bees could still come out of the bottom and do whatever they liked. And by this time their chief pleasure seemed to lie in stinging me.

By the time I got the two halves of the hive into the back of the car, I'd been the unwilling recipient of their close and painful attention no fewer than four times. This would have been okay but for the fact that it was happening while I was actually carrying the hive. It was a novel experience, wanting to drop the boxes of bees on the ground, knowing that to do so would invite further and much harsher punishment.

Now that they were in the back of the vehicle, I could pacify them a bit by turning on the engine. It is one of the great consolations of beekeeping that the vibration of a car's engine will keep them calm. The other small mercy is that when they escape from the hive en route, they cling to the back window to try to get out, and rarely venture forward. Except for the occasional visitor on my leg, driving up the motorway.

By the time I got home, what with all the delays and distractions, it was nearly dark. I had to put the bees in the back garden of a neighbour's house. Steve and Sarah had been curious about the bees for some time, and had virtually begged to play host to a hive when the opportunity arose. All was fine as I gently moved the brood box into position. And then the supers, which were not quite so tightly confined. In fact, they weren't really confined at all.

And just as I was edging quietly past the side of the house, the security light came on. It was there as a helpful guide to illuminate the pathway for visitors and burglars after dark. Most creatures would probably have ignored it. But not the bees. To them, bright light means sun. Sun means daylight. Daylight means work. So as soon as the light came on, the foragers shot up towards it. On this occasion, they would have had a bit of a surprise, as the 'sun' turned out to be about ninety-three million miles closer to Earth than they had anticipated. No further, in fact, than the roof of my neighbour's house.

So the day ended with the hive moved, the inhabitants totally disorientated, the beekeeper stung several times, and a couple of thousand bees clinging to the security light. In the Bad Beekeepers Club, that's what we call a result.

FEBRUARY
INTERLUDE

I've just spent half an hour appearing to do absolutely nothing. To an observer, it would have seemed as if I was mad. Standing at the back of our garage, peering intently at the green moss growing on the shingle, barely moving. I was, though, having the time of my life; watching the bees.

It's mid-February, and the first real signs of life are literally emerging from the hive parked on the front of our flat-roofed garage. The bees which have been surviving the harshness of the winter in a cluster inside are venturing out, looking for sustenance, and they've found it almost on their doorstep in the form of moss. There's no flower there, no pollen or nectar to be drawn, but something they need much more at this time of year: water. The moss is like one large sponge, full of moisture, so the bees come out, just a few at a time, to take the most vital component of their diet back to the hive.

I must admit that I had added to the mix somewhat with an old frame of honey which had been hanging around in the garage through the winter (Bad Beekeeping incident no. 237). When it fell on the floor by the freezer, creating a sticky mess, rather than throw it away (which was the instruction from above), I'd put it on the garage roof to see if any bees would take it at this time of year (Bad Beekeeping incident no. 238). At the height of summer, it would attract a cluster within an hour or two, depending on the scouting abilities of the nearby bees. Today, nothing had happened. It was still too early, I thought, and I would have to chuck the honey away after all.

But when I walked past in mid-afternoon, my eye was caught by a brown cluster beavering away on the yellow honeycomb. They'd found it, and were exploiting it as fast as they could. So I stood and watched them. As it's built into the side of a slope, the roof at this point only comes up to chest height, so I had pretty much an eye-level view of what the bees were doing just a few feet away.

I've seen it happen often before, but it's fascinating every time. The bees are oblivious to their surroundings. They have only one mission in mind – to get the honey out of wherever it is and back to the hive before darkness falls, or someone else finds it. Once or twice they flew so close to my face and ear that I could feel the draught of their wings. Two instincts kick in here. One is to withdraw or even swat them away (and we know what a bad idea that is). The other is simply to stand still and enjoy the moment, for that is all it is, and it will pass all too soon.

Beside the honey gatherers, the water carriers were quietly going about their business. I could see them find a spot of moss and stick their tongues out to suck the moisture away. Actually, the proboscis isn't really a tongue. For a start it's as long as the bee's whole head, and it looks ramrod stiff – like one of those pipes used when jet fighters refuel in mid-air.

When I'm watching the bees like this, it's almost like daydreaming. I'm standing still and quiet, trying to see as much as I can of the detail of what's going on. You can see the characteristics of each individual as it labours away. These bees have a beautiful tangerine colour, but every one looks ever so slightly different. In the background I could hear the blue tits chirruping, and the woodpecker yipping away as it does at this time of the year. But I like to hear it, for it means that there are indeed woodpeckers about, which is a blessing in itself. And it indicates that it's the time of the year for things to start happening again.

After half an hour or so I stir, as if from a slumber, unaware of exactly how long I have spent rooted to one spot. But it has not been time wasted. For I have been taken out of myself, thinking of nothing other than the activity before me. I feel inspired and refreshed. And I can't think of anything else in this life which has the same effect.

Chapter Seven

THRIFT

How parsimony is one of the beekeeper's closest friends; how much it could all cost; and why one should always exercise caution at an auction.

Starting up in beekeeping doesn't necessarily come cheap. True, when you first join your local beekeeping association, they'll lend you stuff. Many associations will let you have a hive and bees for nothing for the first season, while you try it out. Once you've decided to take it on full time, then you can buy the whole set-up and off you go.

But the beginning can be expensive. The list of essential hardware you need is quite extensive:

Hive (stand, floor, brood box and supers, queen excluder, crown board, roof); wooden frames with wax foundation; beesuit with veil; gloves; hive tool; smoker. Total cost new: at least £500. And those are just the basics. Later you can add to this: an eke, mesh floors, extractor, settling tank, uncapping tools, feeders, jars, lids, labels, medicine, a blowtorch, a hand spray, bee brush, skep, straps, honey buckets, a strainer and a wax melter. Total cost: infinite. By now you'll have so much stuff you'll have to build a shed at the bottom of the garden to take it all. Cost, if you don't build it yourself: hundreds.

At this point you'll be quite adept, so you'll need more specialist equipment: frame cleaners, more hive tools, manipulation cloths, gauntlets, floor scrapers, straining bags, honey creamers. It's a gadget collector's dream. Which is why, so my wife claims, I love it so much.

All this is rather ironic as it turns out that beekeepers are among the thriftiest people in the world. Not mean, I must hasten to add. Just thrifty. It's not a rich man's hobby; it doesn't have the right kind of bragging rights. You can't say things like, 'Have you seen my gold-plated WBC Hive? I had it put together by a team of blind Abyssinian carpenters.' You *can* say things like, 'Not a bad honey crop this year really. Just the odd two hundred pounds off the one colony.' But only if you're a good beekeeper. The Bad Beekeeper will say, 'Honey? You had honey this year?'

No, beekeepers are not wealthy people. Like the bees themselves, they keep a close eye on incomings and outgoings, making sure they stay on the right side of Mr Micawber's formula: *Annual income twenty pounds, annual expenditure nineteen nineteen six, result happiness. Annual income twenty pounds, annual expenditure twenty pounds ought and six, result misery.*

Fortunately, you don't have to go out and buy entirely new equipment, apart from the real basics. There's loads of 'previously owned' stuff floating around the world of apiculture; not just second-hand, but third-, fourth- and fifth-hand as well. Some of the bits and bobs I've come across and then tucked away in the shed, to be forgotten for all time, seem to be ancient.

Most associations hold an auction of fellow beekeepers' cast-offs every once in a while, and judicious purchases here can save you a lot of money. On the other hand . . .

The biggest and best auction I've been to is the West Sussex Beekeepers Association event held down at Pulborough every May. They have stacks for sale every year, and it takes hours to go through it all. It's always entertaining though, and a good chance to compare notes. It also provides the Bad Beekeeper with another royal opportunity to cock things up.

The items for sale are usually laid out in long lines on the grass at a local college on a Saturday morning, and as it's springtime they can be coated in a fine drizzle by the time the bidding gets under way. You'll find everything from brood boxes and supers to large plastic bags of sugar, frames, roofs, floors and wax foundation. Plus various lots whose contents are not entirely familiar to anyone present – strange-looking items of home-made gadgetry of unknown purpose, which the new owner will have to puzzle over.

Once the auction is under way, it's all too easy to get carried away in the excitement of it all. Sometimes people get so wound up in the contest of making the winning bid that they end up offering more for something second-hand than they would have paid if it were new.

The Bad Beekeeper can top even that. My first time out at Pulborough, I came across a book which I thought was not only instructive but also very interesting: *A Manual of Beekeeping* by E.B. Wedmore. First published in 1932, and reprinted at regular intervals since then, it is a compendium of invaluable advice and information on beekeeping methods. And there it was, with all the other items on the grass. I stood by it early, to be in a good position to get my bid in. Slowly but surely the gaggle of auctioneers and interested parties made its way round. When the bidding started I raised my hand: two pounds. There was another bid from the other side of the crowd. £2.50. I raised my hand again. Three pounds. And so it went on for a while. I knew the price was rising, but this was a really useful book in good condition, and I didn't know when I might come across it again. I *wanted* it. So I raised my hand again. And again. Distracted for a second, I found the auctioneer pointing at me. I raised my hand yet again. In a tone of high authority and slight impatience, he announced in a clear voice loud enough for everyone to hear, 'Sir, the bid is already with you.'

I had just attempted to bid against myself. D'oh. And yes, people laughed. But I got my book. I paid £8.50 for it, which wasn't too bad, seeing as how it had cost £14.95 on publication. And copies on Amazon, new and used, go for five to ten times that amount these days. So I was pleased with my purchase, until I got home and went to place it on the bookshelf alongside my other works on apiculture, where I found a space next to . . . *A Manual of Beekeeping* by E.B. Wedmore.

Not only had I bid against myself at the auction, I'd also bought a book I already owned.

I think I outdid even that prize moment with my purchase at the same auction the next year. It was a metal box – in essence a cube – with a removable lid and an electric element inside it, which lay underneath a perforated metal platform. The auction programme had it down as a wax melter, though no one was quite certain as to exactly what it was. There was no plug attached to ensure that it was working, and no instructions on how to get it going, if indeed it was in working order.

So of course, I just had to have it. Wax melting is part of the beekeeper's repertoire of winter activities. The bees make a lot of it during the summer months, and when melted down it can be exchanged with suppliers for fresh sheets of foundation. It's wonderful way of recycling. That's why there are lumps of wax all over the place at Turnbull Towers; in the shed, on the garage shelves, tucked away in plastic bags, and occasionally, under the driver's seat of the car.

Good and Bad Beekeepers alike also harbour desires of making candles with their wax. If you're gadget-minded, it's another opportunity to spend lots of money on things like the Kochstar Wax Melter, which with the accompanying dipping station will set you back several hundred quid. And that's before you pick any of the more than 150 different moulds on offer, from the Dumpy Owl to the Eiffel Tower.

I did buy a basic candlemaking kit a few years ago; just a straightforward candle mould with a wick. It's in a cupboard somewhere, waiting for the wax.

You see, the Bad Beekeeper does rather get stuck on the basics, like how to melt the stuff down in the first place. The cheapest and most natural alternative is to make a solar wax melter. It's a big wooden box with a glass lid. You park it in the garden, put all your old frames into it, and wait for the sun to do its job. In Britain, of course, this can take some time. But eventually enough heat is generated for the wax on the frames to melt and ooze out through a filter into the bucket that you hope you've remembered to leave there. Otherwise you end up with a pile of freshly molten wax on your nice, newly mown lawn. Which doesn't go down well in certain quarters.

Good beekeepers are skilled enough to make their own solar wax melters. As I'm totally cack-handed at anything carpenterial, I had to seek help from a kindly old gentleman in our village who made bird boxes in his spare time. I may have given him slightly generous measurements, for the result was the mother of all wax melters. It did everything it was supposed to do, but it weighed a ton; there were complaints from certain sections of the Turnbull community that it spoiled the view of the garden. So for a while now it's been consigned to a dark corner where it fulfils its new function as a general dust-gatherer and leaf collector.

So what I needed was something more compact, something less obtrusive, that would get the job done winter and summer; something that would not offend the eye.

And here, on the damp lawns of a further education college in Sussex, it was. It gave me the sort of thrill that a teenager gets from a brand new games console. But how much was it worth? What would you pay for a stainless steel box with no plug attached?

I reckoned I would go up to twenty pounds. A bit of a gamble, I knew, but it would come in handy for melting all those lumps of wax that had been cluttering up the place for years. Unfortunately, someone else in the crowd had the same idea, and so the stage was set for one of those vicious bidding battles. We started at ten pounds, then twelve, fifteen . . . I was getting close to my limit too soon. But there was one other guy who also seemed very keen, and who kept on bidding against me.

It's amazing how quickly calm and reason can evaporate in such situations. I knew I'd set a cap on my bids. But I'd had my eye on this little beauty for hours, as it sat there glistening gently in the drizzle. I'd stood by it, slowly getting damp, as the auction huddle made its way along the row of lots, painfully slowly. I'd exercised my patience and, damnit, I wasn't going to give way, not now. Twenty pounds. Twenty-two. My adversary looked so calm, so self-assured, even smug; while for my part, irritation and resentment were beginning to take hold. Why would he not give in? At twenty-five pounds, I tried to deliver a knockout punch by putting a full extra fiver on it. That should sort him out. But no, Mr Serenity just kept coming back with more.

By now the little metal box was immaterial. It was just a question of winning. What could the other guy possibly want with this piece of crap? It had to be mine. At last, at long last, but probably laughing inside himself, my opponent gave way. The final bid was mine. 'At forty pounds,' the auctioneer cried with what sounded like mirth in his voice, 'SOLD.' He might have added, 'to the idiot on my left'.

As I revelled for a moment in my triumph, I could detect people around me ever so slightly shaking their heads. 'Forty quid for that piece of junk?' they were thinking.

'What's it do, anyway?' someone said.

'It looks like a hospital autoclave.'

And reality smacked me right between the eyes. I'd bought a steriliser. An NHS cast-off, that had probably been lying unused in a dusty cupboard for years, waiting for a fool and his money to go their separate ways. The joy of victory floated away as I handed over my four tenners. I knew there would be questions at home:

'You paid forty pounds for what, *exactly?*'

I did get it to work, though. Put a plug on it, and the element quickly heated up. Next question: was it a dry or a wet steriliser? A question quickly answered when the element turned red hot and burned out. New element bought and fitted – not by me (don't be daft), but by Steve, my technically proficient neighbour – water added, system worked. So what I had now, I realised, at a total cost of fifty pounds, was a steamer.

Melting wax in it was a slow and difficult process, as the thermostat was impossible to decipher and would either take hours to heat up or would boil over in a trice. After one or two 'little accidents' in the kitchen, I had to accept defeat and consign it to the shed, where it remains to this day. Until the next time I attend the West Sussex Beekeepers auction, when I might have a little something of my own to add to their list of items for sale . . .

One of the joys of beekeeping is that once you know what you're doing – and as I've proved, even when you don't – you can improvise with equipment, and try making your own. Me, I'm a terror for recycling frames. The woodwork costs more than a pound a time for a new one, and all you're paying for is five thin strips of pine, which you have to knock together with little black pins. Much better, in my view, to clean the wax off the old frames and use them again with new foundation, thus saving the pine forests for future generations – or at least a branch or two.

This does, however, require the use of what's called an Easi-steam, which is in essence one of those wallpaper removers converted for use on a super or brood box. Since one of these costs about £70, you've got to recycle a fair number of frames to make it pay for itself. Plus there's the added cost of the electricity to heat it up, and all the hours of labour that you have to put in. In other words, it may be a false economy, which is why it's perfect for the Bad Beekeeper.

But there's more. While a lot of the wax does drop off the frames in the Easi-steam and dribble out through the mouth at the front, there's still a fair bit of gunk attached to the frame, which now has to be knocked apart and scraped off before it can be reassembled with new wax foundation. I like to do this sitting at the table on the patio just outside our kitchen. From there I can enjoy the view down the garden, and still engage in cheery conversation with my beloved as she crafts another fine meal for the dinner table. Occasionally bees, drawn by the scent of the wax, will pay a visit. This might put some people off, but to the beekeeper it's the equivalent of the bluebird alighting on Snow White's outstretched fingers. It's a delightful moment.

The only drawback is that the whole area becomes coated with the wax and propolis that have been meticulously scraped off the frames. And though I try my best to tidy up, some of it does get left on the wooden table, and on the bench, where it gets sat on, and where it then sticks to one's trousers. Wax will eventually come out in the wash. But propolis is another matter. It sticks to anything. For ever.

I must confess I am guilty of embarrassing my bees on occasion. Late in the season, it's important not to have any holes in the sides of the hive where strangers can get in. Otherwise, before you know it, there'll be hordes of bees from elsewhere ganging up to get in and steal the honey inside; honey they can smell from some distance away. Trouble is, these things creep up on you. You don't notice them all summer when it's not important, and then all of a sudden you realise that one poor colony is as vulnerable as an old lady with a bag of fivers in Ali Baba's cave. Quick remedies are called for. But I'm sure the makers of Blu-Tack never imagined that their product would one day be used to plug holes in the corner of a beehive.

Sometimes it's not so much a small hole but a long crack that's letting the invaders in, in which case the gaffer tape comes out. It has an immediate effect, so long as you make sure there are no bees stuck to it when you tape it down. Wise beekeepers will tell you that this is fine, so long as it's a temporary remedy. Recently I found a brittle white substance in the corner of a super, and wondered what it was. Then it dawned on me. It was the remains of a lump of Blu-Tack I'd put in there a year earlier and forgotten about. And one poor hive will never live down the humiliation of having a long grey belt of duct tape round its middle for the whole winter. If not the mother, then necessity is probably the rather wrinkled and stingy maiden aunt of the Bad Beekeeper's sense of invention.

I'm sure we get this sense of thrift from the bees. They don't waste a thing. So nor do we.

Chapter Eight

ENDURANCE

*Wherein the author warns the reader of the
discomforts that may lie ahead; though only if they're
as clumsy as he has been.*

Honey is sweet! But the bee stings.

Proverb

Some beekeepers will tell you that you shouldn't really go on
too much about getting stung, as it might put people off. I
say forewarned is forearmed, and you might as well know
what you're letting yourself in for. Although had I known just
how much I would get stung in the early days, I might never
have become the bad beekeeper I am today.

So here, in no particular order, are the worst stings I've
ever had. And if you are considering taking up the noble art

of apiculture, console yourself with the idea that you'd have to work really hard to get stung as much as this.

I didn't help myself with the notion, during my first solo season, that smoking the bees was disruptive to the hive and threw things out of their natural order. There's a rule written somewhere that every time you smoke a colony, it loses three days' production. So in my naive eagerness, I adopted this as a working practice: not to smoke the bees when performing an inspection.

Now, even the mildest-mannered bees will take exception to having their routine disturbed. How would you feel if the roof of your house was ripped off once a week and the interior was torn apart? Exactly. So when I opened up a hive without the distraction of smoke, even the house bees – normally the more docile members of the colony – would rocket out of the frames and angrily bounce off my veil. Never mind, I thought, it's for the good of the craft that I do this. Until one day . . .

It was the first hive I took to the farm. Not the kindest colony of bees I ever had. In fact it was one of the worst. The beekeeper who gave them to me said they were quite mild. If they were nice, I'd hate to see what his nastier bees were like, for this lot were vicious. I tried to re-queen them once to improve their temperament. This meant taking a gentle, even-tempered queen, putting her in a loo roll with thin paper wrapped around it and sticking it in between a couple of frames. The idea was that although initially the bees would want to kill the new comer, by the time they'd chewed through the paper they would have become used to her pheromones, her scent, and would accept her as their new queen. In this case, her aura cannot have been to their liking, for all I found a week later was a little pile of chewed-up cardboard on the floor of the hive, and no sign of a queen.

Anyway, there I was, on the farm with these badass bees pinging off my veil. A number of them decided to stay there for good measure. Quite a large number, actually. Maybe they wanted to get a closer look at me. I carried on working as best I could until that nightmare sensation – horribly familiar – struck me for the second time in my brief bee-keeping career. As well as the scores of bees walking around on the outside peering in at me, there was also one, just one, yet again, doing a tour of my head *inside* the veil.

You'd think I would have learned, wouldn't you? But I hadn't. I'd done up the zips around my neck, but had omitted to cover my Achilles heel, or in this case, my Adam's apple. The point where the zips meet leaves a gap just big enough for a bee to crawl through. There's a Velcro flap there specifically to cover it. In keeping with the traditions of bad beekeeping, I'd forgotten to do it up. So one of the bees on my veil – unsmoked and rather irritated – had found a warm dark place to explore, and gone inside, onto my neck.

It's a curious sensation. First the moment of realisation:

Oh no, not again. Then the heat as the flush of panic rises up the face. Then the sweat. And bees love the smell of sweat, remember. Then the question . . .

What to do?

Take the veil off.

Can't. Face covered by dozens of not very happy bees.

OK, what then?

Stand still, don't move, and hope it goes away.

And where's it going to go? Do you really think in all that dark, warm humid jungle of clothing and skin and hair, it's going to say to itself, 'time to go home', find the tiny hole it came in through, and leave?

So what do you suggest?

Wait. Something might crop up.

So I waited, and moved away from the hive in the hope that some of the bees clinging to the veil might go home. They didn't. And what 'cropped up' was not exactly advantageous. The invasive bee took a long, leisurely stroll around my head, to the back, where they love to go. There must be a scent gland or something there. Just behind the ear. Ah, the ear, another warm, dark place to explore. Smells interesting, too.

Then I remembered the advice, the warning echoing in my mind like a flashback in a film. If a bee goes inside your ear, you're in real trouble. Once it's in, it can't turn round . . .

By this time, my little friend was traversing the rim of my left ear, and it was time to act. I would have to kill her before she entered the ear canal, and hope that she would die before she had time to sting me.

Some hope. I lifted my gloved left hand, pointing my index finger like a pistol to the side of my head, and jabbed. And again. And again, twisting the finger this time. That bee would by now have to be not just dead, but surely well and truly squished.

If there's one place more painful to be stung on than the head, excluding certain very fragile parts of course, it's probably the ear. It's got cartilage and other sensitive hard stuff and none of that nice fatty tissue to act as a cushion. You remember the hammer blow, right? This was more like a dull spike being banged into the side of the head, through the ear. It hurt. It really hurt.

Step number one in such a situation is to remove the sting. Difficult right now, as I couldn't see exactly where it was, and I still had a faceful of bees on my veil. Remove the veil and I would probably get enough further stings to dissipate the pain of the original, but I wouldn't look very good.

Step number one (a) then: remove bees from face. Retrieve bee brush from back of car (thank heaven I managed to remember it this time). Brush bees furiously from veil. Get in car. Shut door. Unzip veil and try to remove sting with fingernail.

I completed all but the last manoeuvre. I couldn't see the side of my head, and the poison sac at the tip of the sting had had enough time to pump its every last nano-drop into me.

By this point, I was panting, hyperventilating with the pain, which was compounded by my fear of what might happen next as the venom tried to spread through my system. In those days I used to swell up quite a lot when stung (I suspect my body has become used to it these days, because my reaction is far less dramatic). Would my face swell to grotesque proportions? Would my head explode? What should I do next? At that moment I was incapable of anything except sweating and trying to catch my breath, as the pain pounded around my skull.

By the time I got home, the poison was working good and proper. Half of my face blew up, the other half stayed relatively normal. My left ear expanded horribly, giving me the refined look of a village idiot, while large flaps of skin descended beneath my jawline. I had more wattles than a Christmas turkey.

I was so ugly, I didn't even take a photo. My wife was horrified. And the doctor – to whom she despatched me – wasn't very impressed either. But the worst of it was, I had to take a day off work. I wasn't even due on the next day, but twenty-four hours later I was still too misshapen to appear before a TV camera, and my co-presenter Sian Williams had to tell viewers why I wasn't there. That was my only real regret from the whole experience. For I had, after all, learned a lesson or two:

1 Always make sure the Velcro flap is sealed.
2 Smoking bees is actually not a bad idea.

3 Nothing I ever did in beekeeping could surely hurt as
   much as that again.

That last lesson was the most important. I had turned a
corner in my career as a Bad Beekeeper. For whatever disas-
ters lay ahead – and there were plenty still to come – none of
them could surely be as unpleasant as the one I had just been
through. Or so I thought.

They were nasty bees, that's for sure. They also had the
nasty habit of following. That's when one or two of them
decide to stick with you in a rather aggressive manner, even
when you've shut up the hive and walked away. It's unpleasant as it means you can't unveil until you've gone some
distance, even a couple of hundred yards, from the hive. And
once a couple of them have got into the habit, it encourages
others to do the same.

The general advice when you find yourself being followed
is to go and stand under some foliage, so that the bees get fed
up and go home. I have to say it's never worked in my experience. Either I must always have had really persistent bees,
or they find me fascinating company. But I could probably
have taken root before this particular lot would leave me
alone.

At this time I was trying to teach my younger son, Will,
the basic elements of beekeeping. I hardly knew them myself,
but at the age of twelve he was still interested in that sort of
thing, so I took him along. The nasty bees were in their usual

playful mood, so we didn't stay long; just doing what we had to do, closing up and leaving, accompanied by the usual retinue. Perhaps I was being overprotective, for I swished and flapped at them to force them to retreat. This they duly did, and we withdrew to relax for a moment by the back of the car, and took off our veils.

Then they came back for some afters. A whole squadron this time, and looking for action, like dive bombers homing in on a naval convoy. 'Quick, quick, get in the car!' We shut the boot. Will got in the passenger seat, and wound up the window. Fortunately it was me they seemed to be after, so I found myself skipping round the car, madly flapping over my shoulder at the aggressors with one of my gloves. I must have looked a sight – like a demented morris dancer performing solo in the woods:

> Oh I'm a jolly beekeeper, I love to dance and sing.
> I've got these nasty bees on me, and now they want to sting.
> I just want to be friends with them, but every time I pass
> They come out to attack me, and sting me on the —
> Chorus: Feet! Hands! And knobbly knees!
> He's been stung by the devil's bees!

My dignity may have been wounded once again, but this time at least we escaped unscathed.

This was clearly a warning that with these bees, Something Had To Be Done. It was time to re-queen them again. This time, with more success. I wish I could tell you what happened to that particular colony, but I can't. I should have notes to remind me of where they went. But being a Bad Beekeeper, I didn't record their history. Shame.

Later, I took to carrying a useful device called an Aspivenin in case of stings. It's a 'precision-made mini-pump' which works like a syringe in reverse. You press it to the part of the body in need of attention, push in the nozzle, and it creates a vacuum, pulling the skin outwards and sucking out the venom at the same time. It's very useful because it stops the poison entering the bloodstream and causing all that potentially embarrassing swelling.

I once managed to get stung through the veil in the middle of my forehead. I'd been leaning down towards the entrance of the hive to look at something, my head pressed against the mesh that was supposed to be protecting me. One of the guard bees quite reasonably took exception to this and did what it felt it had to do. This time I managed to retreat without a posse of bees on my face, whipped off the veil, and applied the suction syringe. It worked beautifully, removing the venom and the threat of temporary disfigurement. Just as well, as I was due to be on air the next morning with the lovely Natasha Kaplinsky.

There was, however, one drawback. The legacy of sucking out the venom from underneath the skin is a bit of a bruise; you have to pay somehow for getting away with half a sting.

So I awoke the next morning to find a perfectly round black mark, the size of a squash ball, plumb in the middle of my forehead. No matter, I thought, it can be covered by make-up before we go on air, and no one will be any the wiser.

When Natasha walked into the newsroom at five am, I was already sitting at my desk. One look at my face and she knew what I had been up to. One look at her face and I knew that she was looking forward to telling the viewers what I had been up to. It took quite a lot to dissuade her from revealing what was hidden underneath the layers of foundation and powder that had to be applied to cover up my misdemeanour.

To truly excel at the art of bad beekeeping takes a combination of circumstances which does not occur often, like an alignment of the stars. This, mixed with the requisite chain of bad decisions that constitutes the armoury of the BB, form an exquisite melange of ineptitude with deliciously uncomfortable consequences. One such was the day the fish man came to call.

It was September. Never a good time to be messing with the bees, as they are trying to settle for the autumn and reckon, usually correctly, that anyone entering the hive is just trying to nick their stuff. That's what I was doing on the Sunday afternoon with the bees on the garage roof.

I had put them there as a curiosity really, a symbolic guardian for the house, as you had to walk past the garage to get to the front door. At first we had them perched right on the front edge, so that you could stand underneath and get a unique perspective of them flying in and out. As the months wore on, and their mood became less benign, I moved the hive back a bit, so that their flight line would not cross the approach of any visitors. It wouldn't have been fair to the postman.

Summer, such as it was that year, had been and gone. Because the weather had been so inconsistent – in other words persistently wet – I hadn't collected any honey up to that point. Now, on a breezy Sunday afternoon, the rain had stopped for a day, and I felt I had to seize my chance.

It had been a while since I'd harvested, and I was rather rusty at the procedure. In my haste I'd forgotten a crucial step, which was to allow the bees time to move out of the honey supers at their own pace. Good beekeepers do this by inserting what's called a clearing board underneath the super for a while. This lets the bees go down, but not back up again, thus 'clearing' the super. Because it had been raining for so long in previous weeks, I had been unable to get in there and insert the board for the requisite length of time.

So here again is an example of the classic bad beekeeping thought process.

Need to get honey out of beehive. Should I:

a) exploit the dry weather to open the hive, put in a clearing board and wait for another non-rainy day to harvest?

Or,

b) calculate that it's probably going to chuck it down for the next forty days so I might as well steam right ahead and do it all today.

Given the vagaries of our lovely British climate, there is a certain logic to (b), isn't there? The logic of the damned. So, naturally, that was the path I took. And given the weather outlook that day, I had to do it in the one way that beekeepers should never act. In haste.

I'm not proud of what I did next. But don't worry, I got my comeuppance. In my experience, the bees always get the last word.

I resorted to the unorthodox and controversial 'smoke and shake' method to get the honey out. This means taking the hive apart, lifting the honey frames off one by one and removing the bees by smoking, shaking and brushing, then putting them into another super to take away. It's quick and effective, but there probably isn't a better way to wind up a colony. Bad enough to do it on a warm summer's day when the bees are still getting some honey in. But on an overcast autumnal afternoon, with a breeze blowing, it's getting on for suicidal. And looking back, I know I must have been mad.

At one point my wife came out and asked in that kind but concerned way that wives often do when their husbands have gone temporarily insane, 'What are you doing, darling?'

'Just harvesting the honey, my love. Nothing to worry about.'

I knew she wasn't convinced, because she just stood there and looked up at me. My wife may not know a lot about bee-keeping, but she does recognise madness when she sees it. Think of it in the same tone as:

'Why are you carrying that shotgun, darling?'

'Just going to polish the trigger, sweetheart. Nothing to worry about.'

Exit wife, preparing for call to emergency services.

First, I had to crack the hive apart. They'd already started gluing it up for the autumn, filling in all the cracks with propolis to make it draught-proof for the winter. So, quite reasonably, the bees were irritated. Think how you'd feel if someone came along and started ripping out all your double glazing.

Actually, they weren't just angry, they were incandescent, bouncing off my veil with particular vigour. Time and again I drew a frame out of the supers, smoked it (which produced a vicious hissing sound) and shook the bees off, back into the hive, where they joined a throng of growing outrage. I make no bones about it now, it was ugly. Yet somehow I got through the operation without being stung, and withdrew with my inglorious haul to the house. But the bees were not done yet and went for anyone going near them that day. They were so hopping mad that they sent patrols out looking for trouble. Early that evening I went to the chicken run to check on the hens, and was buzzed by an angry pair even then, on the other side of the house where they never normally fly.

Still, I had the honey, didn't I?

The next morning they seemed to have settled. It was, ironically, a sunny and dry day – the one I hadn't expected. So I spent it digging a large hole in the garden in order to plant a tree. Not the sort of thing I usually do in my spare time, but I'd inherited it from a neighbour's garden and it needed to be planted. It was a long, sweaty, rather dirty job that took most of the afternoon ... until Phil the mobile fishmonger turned up.

Every Monday, Phil would come to call in his blue van, having driven down from Grimsby to offer the housewives of Buckinghamshire – and me – his finest fish. He would park his van up the driveway and our transaction would be completed fairly swiftly; I'd have been given my instructions on which fish to buy. We'd compare notes about the weekend's football results, Phil being a long-suffering Grimsby supporter. Cash and cod would be exchanged, and he would drive away.

So as usual, Phil parked a couple of yards short of the

garage and opened up the back of the van. Even though the contents were well iced, the aroma of fresh fish and seafood would waft up as the door was lifted. Which would have been fine, but for the fact that the wafting was also reaching the beehive on the garage roof, and alerting the inhabitants to the fact that something was up, down below.

Had I known what they had in mind, I would have asked Phil to park a little way down the drive. But they'd put up with his presence every week without complaining, so it hadn't been a problem. Until now.

Perhaps it was the scent of the sea bass that tipped them over the edge after the outrage perpetrated on them the day before, for they came down with venom, like Stukas, determined to have their revenge. Curiously, they didn't go for Phil, but latched straight on to me. Now that I think about it, I was wearing the perfect lotion to drive bees wild: sweat and dirt. They simply hate the smell. And at that moment, I reeked of it.

On the first raid they buzzed around my face, hovering angrily around my nose and the back of my head. I tried to tough it out, to save face in front of Phil. After all, these were my bees; I was supposed to be the one in charge. Eventually though, I had to withdraw with some alacrity, handing Phil the money and racing off round the side of the house with the fish. Exiting, if you like, pursued by a bee. Or three.

I waited for a minute or two, but I had to come back for the change. When I returned, Phil was still there, cool as a cucumber and as calm as Clint Eastwood, and the bees seemed to have withdrawn. But only, it turned out, until I reappeared, and then back they came. They'd been lying in wait, and this time they got their man, fair and square. On the eyelid. Again, I had to beat not so much a hasty retreat as a hell-for-leather dash for cover. Dignity be damned; if I hung around any longer I thought they might get the other eye as well.

I felt a little bit guilty afterwards for having left Phil there on his own, but I needn't have worried. He just packed up his van and drove away as if nothing had happened. It was my vanity that was wounded more than anything else. Of course, it hurt like hell. Try jabbing a needle into your eyelid and you'll see what I mean. But the real problem was that my face swelled up so badly that I couldn't go out in public for at least a day.

I had known that the bees would be in a bit of a mood, but what I couldn't work out was why they had waited until that particular moment to strike. After all, I'd been out in the garden digging all day and would have made a perfect target. But they'd waited until Phil turned up with his van. When later I related this conundrum to Michael Young, my bee-keeping friend from Northern Ireland, his answer was simple. 'Didn't you know?' he said. 'Bees *hate* the smell of fish.'

Sometimes you do need a thick skin to carry on when you've been stung for the umpteenth time. But in my experience, the longer you keep bees the less likely they are to sting you, as you learn over the years how to avoid upsetting them. As the Buddhist saying goes:

'Endurance is one of the most difficult disciplines, but it is to the one who endures that the final victory comes.'

And you know what? It feels really good when the pain wears off . . .

APRIL FOOL

April is the month when the Bad Beekeeper's mistakes really find him out. The bees, increasing rapidly in number, expand and explore the hive, much of which has been empty and unused through the dark cold months of winter.

When a bee finds some space, it can't bear to leave it alone. It must be filled up. It's something they just have to do. Some women just have to go shopping. Some women just have to clean their homes relentlessly. And bees, who are overwhelmingly female, are driven to fill their homes with wax – thousands and thousands of hexagonal cells made of the stuff, into which they put food in the form of honey and pollen, and eggs laid by their queen.

It's fascinating to see how they do it. When you open up the hive you can sometimes see the bees tangled together, forming a chain across an empty space. It looks rather uncomfortable, as if they're caught in a strand of spider's web. But they're making wax, which comes from glands in their abdomen.

The good beekeeper keeps things under control by making sure that the hive has no empty spaces for the bees to fill up. The good beekeeper does this by checking that there are enough frames of foundation – sheets made of wax – so that the bees can draw the cells out, and everything is in good order.

The Bad Beekeeper omitted to do this at the end of last summer. And then didn't check early enough this spring. He forgot that there were empty spaces where he'd meant to put in more frames, but the weather got in the way or there was some other distraction like a self-assessment tax return. So when the Bad Beekeeper opens up for the first inspection of the year, what a lovely surprise is in store. A labyrinth of beautiful fresh wax, containing eggs, larvae, sealed brood, honey and pollen all mixed together, and created so intricately that as soon as you lift the top board off, parts of it break off and honey drips everywhere.

Naturally, this is what has happened to me. Several times. And the trouble is, you can't just leave them to their own devices as it's impossible to carry out a proper inspection of what we call brace comb. You can't find the queen, you can't see whether they're diseased or not, and you can't get any honey out either. I did leave a colony like this for a whole year once, reckoning that it was the way the bees wanted to live, it was their mess, and so they should be left to get on with it. And in olden times, by which we mean more than fifteen years ago, it would probably have worked. But that was before the advent of the varroa mite. Eventually I could see that the little red devils had struck, and I had to get in there.

It's not pretty. It requires a fair amount of determination and a sharp knife. What you have to do is cut it all out, and let the bees resettle elsewhere. Good beekeepers try to tie it all back into frames with bits of wire. Bad Beekeepers think about that for a second and then decide it's far too ambitious, and that the bees will sort it all out for themselves one way or another so we should leave them to it. Either way, it's a very sticky business. And by the time it's finished, you're left with an unholy mess and tens of thousands of very confused bees. During the great carve-up, large pieces of comb can drop off and crush the bees below, causing carnage. And with a bit of bad luck, you'll also kill your queen.

I know what you're wondering. Of course I did.

Chapter Nine

STUDY OR
HOW TO RUIN
A SUMMER

There's an awful lot to learn if you're going to be a good bee-keeper. There are more than a dozen books on the market at the moment, all of which promise to teach you how to do it. Wisely, in my view, you're reading the one which tells you how *not* to do it.

And if study and taking exams is your favourite pastime, then the world of apiculture is definitely the place for you. The British Beekeepers' Association has a programme of qualifications to keep you up to your eyeballs in textbooks for years: starting with the Basic Assessment, moving on to the General Certificate in Beekeeping Husbandry, the Microscopy Certificate, Intermediate Theory, and finally the

Advanced Certificate. In between these various levels are a whole raft of modules to choose from, including honeybee management; products and forage; biology; pests, disease and poisons; bee behaviour; breeding and history.

And at the end of the road you get to be either a Master Beekeeper or a Show Judge.

Of course, as a Bad Beekeeper, I've completed none of these qualifications. I did get a Beginner's Certificate from my local association after my starter's course, but that was only because they wouldn't have let me loose with bees if I didn't. Since then, there has been no academic progress. Like most people, I just don't like exams. Even now, more than thirty years after leaving school, I still have dreams that I'm about to take my A-levels and haven't done any revision.

The BBKA keeps trying to get more of us to take the Basic Certificate, which is rather like the driving test for bee-keepers. Its syllabus contains a daunting list of more than sixty items of which the successful candidate must be aware. Most of it is common beekeeping sense. But if you're allergic to exams – and who doesn't feel a bit queasy about them? – you're unlikely to put yourself through one again, unless you're particularly keen. I was happy to leave the hallowed halls of apicultural academe to others. Then fate popped up with something that made me study harder than for any BBKA exam.

Like almost all of us, I'd watched *Mastermind* from time to time since its early days with Magnus Magnusson. The famous black leather chair, the darkened set, the ominous and terrifying theme tune of doom-laden trumpets, and the famous mantra: 'I've started, so I'll finish.' Not to mention, of course, the one word which passed into common usage whenever someone couldn't answer a question: 'Pass . . .'

I never imagined, when watching the programme over the years, that I would one day get to sit in the spotlight and face two minutes of questions on my chosen topic. But such are the privileges of being a 'TV personality', that on occasion the invitation comes through to do such a thing.

So when the call came to take part in an edition of *Celebrity Mastermind*, I thought, Why not? I'd seen others of my background do it without making fools of themselves, and it seemed relatively risk-free; unless you're a former minister of Her Majesty's government, in which case the whole country howls at your shortcomings in general knowledge, as in the cases of David Blunkett and Lord Lammy.

So, the producers wondered, what would I like to choose as my specialist subject? Without hesitation, I thought 'Beekeeping'. It was something I knew a bit about (I thought), a subject close to my heart, and the additional study required would add to my overall knowledge. A win–win situation, surely?

It was only after we had nailed down the topic that I realised I had made a potentially enormous mistake. For if

I failed, what was at risk was not just national humiliation in front of an audience of millions, but also any standing I had with my fellow beekeepers. They'd be out there watching, and tutting. 'That feller Turnbull . . . always going on about his bees on the telly, but he doesn't know much, does he?'

There was only one thing for it. To study.

The invitation had come through in early summer. The transmission wouldn't be until Christmas, but the recording would take place in September. I had three months to get ready. Three months to ruin a summer.

How to set about studying the subject? Well, there's been some controversy in the past about how much help the producers give contestants, but in my experience it was both fair and reasonable. They find a reference work which covers your topic, from which most of the questions will be taken, and it's up to you to read it from top to bottom. We had a complication in my case, as the two thin volumes they sent me weren't really suitable. For a start they were reprints from a long time ago, with references still to pounds, shillings and pence. A lot of the information they contained was out of date. So I suggested that we refer to the book that is considered by many to be the bible of modern beekeeping: the *Guide to Bees and Honey*, by Ted Hooper. This was not a bodyswerve, though, far from it, for *B&H* is far longer and more detailed than what had previously been prescribed. And all I had to do was learn all 262 condensely typed pages of it off by heart; for *Mastermind* is, in the first round at least, a matter of recall.

I found that as I read the book, I already knew quite a lot of it, which was reassuring. Less comforting was finding out how much I didn't know, of which there was a whole heap. And how to test my knowledge? I turned to my two best sources – Christopher Beale, my mentor, and Michael Young, my highly knowledgeable beekeeping friend from Northern Ireland. Very helpfully, they compiled lists of possible questions and answers, which I tried to knock into my brain.

It was all straightforward enough, but for one other element: the fear factor. All summer I had in the back of my mind one question: 'What if?' What if the questions are too difficult? What if I have a mental block? What if I can't even remember my name? It's happened to me before, after a fashion.

Every day on the *Breakfast* sofa we start the programme by introducing ourselves. Half of the time, it's my turn to say, 'Hello, this is *Breakfast* with Sian Williams and Bill Turnbull.' It's not difficult. I've done it hundreds of times. It's written there on the autocue. But . . . Convention and courtesy require that as we say our fellow presenter's name, we turn and look at them. It gives the impression that we know and like each other, which is a good thing, because we do.

One morning, however, I must have been distracted by something. I read the intro in the usual way, turned to look at Sian – and stopped. Without the guiding line of script in front of me, I had forgotten her name.

It was only a fraction of a second. But it's amazing sometimes how much you can think in such a short space of time:

You've forgotten her name. What's her name? Idiot. Look at her, you've only known her since 1992. You must be able to remember what she's called.

Eventually I got there. Sian Williams. Of course. And even though it only lasted for a nanosecond, being the ultra-perceptive person that she is, she'd noticed. When you've presented a thousand programmes alongside the same partner, you can develop an uncanny sense of what they're thinking. As the *Breakfast* theme played and the camera held on that wide shot of the studio, we looked as we always do, smiling into each other's eyes, and she muttered behind her teeth, 'You forgot who I was, didn't you?'

So if I could misplace my TV partner's name, even with the aid of autocue, anything was possible.

The day arrived, a Saturday afternoon, and we drove to Television Centre for the recording. Having spent the best part of twenty years working there, it's a building I know very well. But that day, as we were taken down to the Green Room in the basement, where the guests and their families were assembled, it felt like a completely alien environment; rather like the first day at a new school.

There were, as usual, three other contestants: Jan Ravens, whom I already knew from our separate *Strictly Come Dancing* experiences, the comedian Ade Edmondson, and the Rastafarian poet Benjamin Zephaniah. I didn't see them as competitors, though. We were all in it together, the real competition was with ourselves.

When I heard what the others had as their specialist topics, my heart sank, for they all seemed relatively risk-free. Jan had chosen Daphne du Maurier. Ade picked the Sex Pistols 1975–85. And Benjamin, fairly logically, the life of Bob Marley. When it came to my own subject, all I could think was, Stupid, stupid, stupid boy.

Part of me was also wondering how I had come to be here. The other three were all famous for valid reasons: Jan for her hilarious impressions of Fiona Bruce, Anne Robinson and a host of others in *Dead Ringers*; Ade for a long career in comedy, and now acting. I'd seen him nutting the microphone in stand-up nearly thirty years previously, alongside Rik Mayall. His role in *The Young Ones* in the 1980s had been cutting edge and was now vintage TV. And who was that with him? Only his wife. Only Jennifer Saunders. Only one of the funniest women in the Western World. And as one of Britain's 'top fifty post-war writers', Benjamin had a body of work as long as your arm, and an international reputation.

So who was I alongside them? Just a journalist, a reporter turned presenter who'd done a bit of dancing and was now bluffing it as a beekeeper. As we sipped our tepid drinks from plastic glasses in the airless, windowless confines of the Green Room I had, I considered, no place being there.

But there was no backing down now. I'd signed the piece of paper and promised to be there. They couldn't make a programme with only three contestants. So, like a condemned man walking to the gallows, I went through the motions. Going to make-up, listening to the producers' briefing, meeting the presenter (the magisterial John Humphrys), kissing my wife one last time for luck, sitting down on the set . . .

Sitting down on the set? Before we knew it, we were there already. It felt rather like a gallows, with its black, wooden walkway to the place of execution/inquisition. And the chair, the focus of attention. All they had to do was put a hood over your head, slip a noose quickly around your neck and the scene would be complete.

Ade was up first. The Sex Pistols. If you were around in the seventies you already knew most of the subject, there'd been so much press attention on them. As the questions flew in, Ade answered them with ease, and so did I. I knew that. Yep, knew that one, too. Oh, that's easy. Why didn't I pick a subject like this? Like the Beatles, or the life of Ken Dodd; something simple?

Next up was Jan on Daphne du Maurier. Someone I knew virtually nothing about. But Jan, it seemed, knew everything. Names of houses, key individuals, dates, probably phone numbers and passcodes too. She was good. Very good. Which made me feel bad. Very bad.

Have you ever had a moment when, faced with a difficulty, your instinct is to flee? It can happen to a lot of people, especially in live broadcasting. At least I hope it does, because it happens to me. And right now it was happening with a vengeance. *Why why why why why did you agree to do this? Stupidstupidstupidstupidstupid.* That sort of thing.

'And the next contestant, please.'

That's you. Move. Walk.

It's about ten steps. Just enough time for me to think: This set's just made of plywood. A legendary BBC programme and it's knocked together with one-inch ply.

'Your name, please.'

Bill Turnbull.

I sat upright, forward. I thought if the body looked eager, perhaps the brain would be, too. You're in close-up, remember. Don't twitch. Don't blink hopelessly. Don't burst into tears.

'Occupation?'

Mmm, well what are you exactly? 'Chancer' sprang to mind, but I think I came up with TV Presenter or some such. That's what it says on Wikipedia, anyway.

'And your chosen topic?'

Beekeeping. And how to humiliate oneself in public.

'You have two minutes, starting from . . .' A pause, for effect, as if raising a pistol . . . 'Now. What is the name given . . .'

Yes! I know that. One.

I'd thought it was about instant recall, but the questions were phrased like a bowler with a long-run up, so that you could see which sort of area they were coming from and your mind could rush to the right filing cabinet, pull out the drawer, rifle through the folders and scan the documents for the correct answer, all in an instant.

The questions came in thick and fast. No time to think. Just answer.

What is the . . . ? Ping. Yes.

How many . . . ? Biff. Yes.

Then the googly. The one I didn't know. Guess or pass? Don't think, just one or the other. Guess. Wrong. Move on.

Halfway through I did give myself time to think that it wasn't going too badly. Just to calm myself down. And on . . .

Finally, the peep-peep-peep of the whistle. Has Humphrys started another question? If so, I get a bit more time to consider the answer. Phew. No really, phew. I had no idea what my score was. I just knew I hadn't disgraced myself. I could still hold my head up high at the National Honey Show. Sixteen.

The general knowledge round didn't go quite so well. I found out too late that taking a wild guess was a mistake, as Humphrys then had to read out the correct answer at the time, eating up valuable seconds. It turned into a shootout between Ravens and Edmondson, with Jan coming in first.

I ended up with twenty-seven – a score which they said afterwards was good enough to win on any other day. Yeah, right.

At the end of the day though we went home reasonably satisfied. Humiliation avoided, a bit of money raised for charity, another word put in for the noble art. Chalk one up for the Bad Beekeeper.

And never, ever, do that again.

Chapter Ten

BAD BEEKEEPING:
A BRIEF HISTORY

*In which we learn that we are
by no means the first beekeeper to have
messed things up on a grand scale,
and certainly not the last.*

Let's get things straight. Bees have been around for more than one hundred million years. Humans? Depends who you're talking about. Hominids – human-like species – have been on earth somewhere between four and six million years. Our particular brand of *Homo sapiens*? 100,000. So it's really the bees who should be keeping us, rather than the other way round. You'd think, given the huge head start they had on us, that they'd have sorted it out by now. But, sadly for them, they didn't. Instead, they let us creep up behind them and they've been paying the price ever since.

The first people to exploit the poor bees were so pleased with what they'd done that they drew pictures about it. You can see them on the wall of a cave found in eastern Spain, dating back somewhere between eight and eleven thousand years.

The picture shows a figure of a woman dangling precariously from a rope or vine of some kind, reaching into a huge bees' nest with one arm, and carrying a bucket in the other, while some large and angry-looking bees hover around her.

It's almost like an advert: 'At Ug's Honey Hunters we'll go to any lengths to get the best for you.' Or, 'All because the lady loves . . . that funny sticky stuff we find in caves.'

It's a practice still performed today in some parts of the world, including Nepal and southern India. The honey hunters climb down more than two hundred feet over a ledge on bamboo ladders, creating vast clouds of smoke to ward the bees away. They then launch lumps of grass and moss on ropes which stick to the comb like primitive grappling irons, before pulling themselves in to cut away the honey and wax. It's basic but effective. And it looks highly dangerous.

The first people to get organised with bees were the ancient Egyptians. They made cylindrical hives out of clay, and stacked them on top of each other. When it came to harvesting, they would blow smoke in at the back, the bees would all fly out through the front, and the hive would be more or less clear for them to take out the honey. The same technology is still being used in parts of Egypt today, four-and-a-half thousand years later.

Actually, the Egyptians were pretty good beekeepers really. They practised on a commercial basis, and developed honey as a valued delicacy, a sacrifice to the gods, and a medicine for applying to wounds. On the basis that, kept in the right conditions, it never went off (honey has been found in the Pyramids in edible condition), it was also recommended as a preservative. Although by today's standards, the recipe found in one ancient manuscript was rather controversial:

> *Find a ruddy-haired man and feed him til he is thirty years old. Then drown him in a stone vessel filled with honey and drugs and seal up the vessel. When it is opened after the lapse of one hundred and fifty years, the honey will have turned the body into a mummy.*

Clearly, being a redhead in Egyptian times was not without its hazards. There is another old story of some Egyptians exploring graves, hundreds of years ago, and coming across a tightly sealed jar. On opening it, they found that it was full of honey. They duly started to tuck in, dipping bread into the jar and enjoying their impromptu feast, until one of them found a hair entwined around his finger. Further investigation revealed the corpse of a small child, 'with all its limbs complete and in a good state of preservation; it was well-dressed, and had upon it numerous ornaments.' Mmmm . . .

While the Egyptians gave us a picture (literally) of their bee-keeping prowess, it was left to the Greeks to develop the first theories about apiculture. And on occasion, they managed to get it spectacularly wrong. Take Aristotle, for instance. A very clever bloke; he may have brought deductive logic to the world, but when it came to bees, he was a bit dodgy. First off, he didn't recognise a queen when he saw one. And like others at the time, he reckoned that honey more or less grew on trees:

The honeycomb is made from flowers and the materials for the wax they gather from the resinous gum of trees, while honey is distilled from dew and is deposited chiefly at the raisings of the constellations or when a rainbow is in the sky.

He also subscribed to some very strange notions about reproduction. Bee babies, it seemed, also grew on trees:

Some affirm that bees neither copulate nor give birth to young, that they fetch their young. And some say that they fetch their young from the flower of the callyntrum; others assert that they bring them from the flower of the reed, others, from the flower of the olive.

To be fair, Aristotle was quite busy at the time becoming one of the great figures of western philosophy, as well as writing about physics, poetry, music and politics. His *History of Animals*, from which the above quote is drawn, was an encyclopædic description of pretty much every creature known to man. The great man had a lot to write about, so a couple of inaccuracies are understandable.

What the Roman poet Virgil's excuse is, though, is anyone's guess. Like everyone else at the time, he did not allow for the existence of a queen in the hive. The place simply had to be run by a man:

> *He is the lord*
> *Of all their labour; him with awful eye*
> *They reverence, and with murmuring throngs*
>   *surround,*
> *In crowds attend, oft shoulder him on high,*
> *Or with their bodies shield him in the fight,*
> *And seek through showering wounds a glorious death.*

What Virgil and others were doing was primitive observation, looking at the bees flying around and then taking a guess as to what it all meant. For clearly he'd never seen the inside of a hive. He also believed that bees went to battle:

> *Then in hot haste they muster, then flash wings,*
> *Sharpen their pointed beaks and knit their thews,*
> *And round the king, even to his royal tent,*
> *Throng rallying, and with shouts defy the foe.*

All of which sounds remarkably like a swarm, which is ironically when bees look most fierce but are really at their most placid.

Virgil's *Georgics* were in effect a Latin version of How to Run a Farm and Keep Bees, set to poetry. And he did have some sensible notions about the siting of beehives, recommending that they be given 'a settled sure abode' out of the wind, which is correct; away from sheep and heifers who might knock them over; and out of range of noise and unpleasant odours:

> *But near their home let neither yew-tree grow,*
> *Nor reddening crabs be roasted, and mistrust*
> *Deep marish-ground and mire with noisome smell*

I should have read the *Georgics* before I let the fishmonger come to call. Perhaps I should try my own guide to beekeeping in poetic form:

> *If the fish man's bell doth ring,*
> *Be careful, for the bees might sting.*
> *If they sting you on the head,*
> *Take care you don't end up in bed.*
> *And if you want to have lots of money,*
> *Don't think you'll get it making honey.*

It was not until 1586 that anyone formally recognised that the 'king' was in fact a queen who laid eggs. In his *Tratado breve de la cultivación y cura de las colmenas*, the Spanish scientist Luis Mendez de Torres recognised that one key female figure was responsible for the birth of all the bees in the hive – workers, drones, and of course new queens. He still somehow failed to realise that queens could mate.

But now at least the ball was rolling. Charles Butler's *Feminine Monarchie*, published in 1609, was the first full-length book on beekeeping written in English. He had worked out that bees produced wax from their own bodies, and not from plants. He identified the drones as male bees and explained how the emergence of a swarm could be predicted from the tone of the bees' buzzing. In short, once they reached a high C, prepare for take-off. Like all bee authors before him, he made a mistake by asserting that worker bees lay eggs. They do sometimes, but only in rare circumstances and when there is no queen.

Leaps and bounds were made in the late eighteenth century by a Swiss beekeeper called François Huber. Over twenty years of observations, he established the mating process; how the queen is inseminated by several drones on a mating flight at some altitude and distance from the hive. Huber gets the credit for these discoveries, although the hard work was done by his manservant, François Burnens. For Huber succumbed to blindness while still a young man. He sat by the hives and issued instructions while Burnens stuck his hands in and got stung hundreds, if not thousands, of times.

It's been suggested in certain quarters that Burnens may have been putting those same sting-swollen hands somewhere else while Huber was sitting listening quietly to the bees. In her excellent novel *The Beekeeper's Pupil*, Sara George postulates that the young and very attractive Madame Huber may have harboured a passion for the muscular manservant. In his (fictional) diary, Burnens recalls how, out of sight and earshot of the master of the house, he and the lady of the manoir would share brief conversations ripe with delicate symbolism, hinting that the bees weren't the only ones thinking about insemination. In spite of her longings and meaningful glances, Burnens records that he remains faithful to his master and resists the temptation to mate with his own queen.

But that's just one version. On the other hand, he might have been a very naughty beekeeper indeed. All that time watching the drones, recording details of their nuptial flights, might have put him in mind of a dalliance of his own.

Huber's other big discovery was that in the hive there was a specific and regular space left by the bees between the wax combs, so that they would have room to move. He didn't quite cotton on to it at the time, but this was crucial to the development of modern beekeeping. He also developed the Leaf Hive, where the frames of the structure could be examined like the pages of a book. However, the combs could not be removed, so there was no practical purpose other than for observation.

Throughout this period, as for centuries before, honey was being produced in the same old smash-and-grab way. The bees were kept in skeps, and when harvest time came, the whole colony including the queen was killed, and the honey and wax inside removed and processed. It was, as Thomas Wildman observed in 1770, an 'inhuman and impolitic slaughter of the bees'. Wax, larvae, eggs and honey would all be mixed together in a crude pulp, before the honey was sieved off. Wildman worked out how to introduce wooden bars to the skep and move the bees from one container to another, avoiding the need to kill them all come harvest time.

The man who put it all together and changed the world of beekeeping was the Reverend Lorenzo Lorraine Langstroth, a Congregationalist minister from Pennsylvania, who calculated the width of the bee space or channel through which the bees would pass without filling it up. Too small and they would line it with propolis, too big and they'd fill it with wax comb. The magic measurement was between a quarter of an inch and three-eighths of an inch. Once this was established, Langstroth could design wooden frames which could be removed and replaced without breaking up the colony's internal structure. He came up with the world's first movable-frame beehive.

The new design meant that the bees could be inspected and checked for disease. Swarming could be controlled. Weaker colonies could be strengthened with bees from other hives. And best of all, boxes could be mounted on top of one another, revolutionising the production of honey. If the queen was kept in the bottom by means of a queen excluder, only workers could go in the boxes above, and only honey would be kept there. The whole process became simpler and cleaner.

No wonder, then, that Langstroth became known as the father of modern beekeeping. His design has remained pretty much unchanged since he came up with it in 1852. Sadly for him, he never profited properly from his discovery. Although he patented the idea, it was widely copied while he was wading through the laboursome patenting process. He never made the fortune he deserved from such a remarkable idea. But in his lifetime he won fame and admiration, and among beekeepers today, immortality.

The beehive that most non-beekeepers are familiar with, and which most modern beekeepers spurn, is the WBC, invented by William Broughton Carr in 1890. It's the one with the lovely shape that has come to symbolise the beauty of the English country garden. It has a double wall; the inside is in effect a normal hive, encased by the splayed pyramid-style 'lifts', which stack up on the outside. It is said to be good for the bees, as it keeps them cool in the summer and warm in the winter. But they're not particularly popular with bee-keepers, as they require more work to take them apart, and are difficult to move. For simplicity's sake, most of us stick to a single-walled hive with just the boxes containing the frames inside.

What is remarkable, really, is that we are still using the same fundamental hive design that Langstroth came up with more than 150 years ago, before the American Civil War, a conflict fought with cannon, cavalry and muskets. Since then, the rest of the world has moved on to electricity, the internal combustion engine, powered flight, nuclear power and the Fender Stratocaster. In beekeeping terms, we're still stuck with the horse and cart. You'd have thought that by now someone, somewhere, would have had a better idea. All you need is a system that separates the queen and her eggs from the honey factory, and a way of inspecting them safely from time to time. Simple, isn't it? Answers on a postcard, please.

So much for the beekeepers, good and bad, of the past; these days there are some shockers around to give them a run for their money. I've never thought of my bees being worth anything in financial terms. But since the shortage of bees hit home in Europe, they've become a lot more valuable. Prices for small colonies have gone up to £400 on occasion. Add honey to that, and you've got a colony worth quite a lot. And where there's value, theft is never far away. So now bee rustling has started here in Britain. Colonies have been stolen lock, stock and beehive all over the place, from North

Yorkshire and Shropshire, to Norfolk and the New Forest. In Scotland, eleven hives due to be shipped to the royal family's Balmoral Estate were stolen from West Lothian. Stealing property bound for the Crown? Now that must surely still be a capital offence.

What makes such thefts even worse is that they must have been carried out by fellow beekeepers of some kind. Nicking hives isn't like shoplifting. You can't just pick up a bag of bees and slip them into your pocket; you'd have an interesting time if you did. No, it requires specialist technical skills, and, in the case of eleven hives, a fair amount of strength and organisation. Once you know how though, it's probably straightforward. A lot of apiaries are deliberately sited in quiet, remote locations. You can't exactly padlock a hive to the ground. And once your bees have gone, they're not easy to identify. You can't go round to a suspect's apiary and try to whistle them home. Nice idea, though.

Still, at least beekeepers don't go round poisoning people, do they? Well, only occasionally. And in New Zealand. One beekeeper from a place called Whangamata managed to give twenty-one of his customers rather more than they bargained for by selling them honey that contained high levels of a toxin called tutin. Accidently. Symptoms include vomiting, delirium, giddiness, increased excitability, stupor, coma and violent convulsions. You get this when the bees feed on so-called 'honeydew', secreted from tiny sap-sucking vine-hopper insects which feed on the tutu plant. Good beekeepers are supposed to remove their hives from areas where the honey-dew is abundant. The defendant in this case clearly wasn't one of those. To be fair, the first person he poisoned was himself. He spent three days in hospital but no one identified the honey as the cause of his problems. So when he got out, he went home and carried on selling it. After the court case, where he was ordered to pay compensation to most of his victims, he declared his intention to carry on procuring and selling honey. Think of the advertising slogan: 'Buy my honey – so much more than a taste of nature's sweetness. It's Tu Tu delicious!'

Still, at least beekeepers don't go around killing each other, do they? Well, only occasionally. And in Australia. In 2009, Donald Robert Alcock was convicted of murdering Tony Knight. Alcock, in dire financial straits, crept into the victim's home and shot him in the back with a hunting rifle while he was asleep, before making off with £22,000 worth of honey. He loaded the largest tubs onto his truck and drove them to a honey distributor. Then fate played its part. Somehow, while unloading a huge tub weighing 3,000 pounds, he managed to get himself pinned under it, requiring urgent medical attention and a trip to hospital. Queensland police were called to the incident, and when they later discovered the decomposing body of the poor victim in his bed, they put two and two together. Donald Robert Alcock – a very Bad Beekeeper indeed.

Chapter Eleven

TRUTHFULNESS

*A grave chapter, wherein we recall
some truly horrible misdemeanours, and
fear the reader may not like us quite
so much afterwards.*

At this stage, I feel a bit like a young man who has been wooing a fair maiden for a while, and having won her trust and affection, has to sit her down on a park bench and say, 'Mildred, there's something I have to tell you . . .' And she looks lovingly in his eyes and says, 'What is it, Jasper, dear? Whatever it is, it doesn't matter. Nothing could come between us now.' And he goes on to tell her that he's not a viscount after all, and he was the one who stole the collection plate at school and got her brother expelled, and that it was his stupid prank that caused the pony and trap acci-

169

dent which triggered her father's stroke, and it was his best friend who'd got her sister 'into trouble' and he'd known about it all along and covered it up. And so on and so forth.

But enough about my brother. Because, dear reader, there's something I have to tell you. I've been putting it off, chapter after chapter, almost hoping that our relationship wouldn't get this far. (Unless you've skipped halfway through the book, which isn't fair.) But, damnit, it's time you knew the truth. So here goes.

It was May 2001, the twelfth to be precise, when I first stumbled into Robin, the other beekeeper in the village. Stumbled was indeed the right word, as it was Cup Final day, and I was wending my way home after a long beer-drenched afternoon in a friend's garden. Robin was on his way home from the bees, something I cleverly deduced from the fact that he had his protective suit and veil on, and was carrying a smoker. We struck up a conversation about the noble art, and as this was my first season he kindly invited me to help him look after his bees. He was now in his seventies and suffering from Parkinson's disease, and he needed some assistance.

I, of course, was more than happy to oblige, and that summer went with him almost every Sunday afternoon to check his two hives, which were thriving. He'd been beekeeping for twenty years and yet, with characteristic modesty, still claimed to be a novice. He taught me a great deal, especially about patience. I had a tendency to rush things in those days, to make rash decisions instead of taking in the full picture and carefully considering the options. I helped him lift the heavy boxes on and off the hives, and come harvest time, took off the honey and bottled it with him. He used to get even more wasps in his kitchen than I do in mine. But here I learnt that even a wasp will leave you alone if the atmosphere is thick enough with honey.

So all was going well. Until one day . . .

Towards the end of the season we'd made our regular inspection of the bees and were closing up one of the hives. Everything was in good order as we started putting the frames back in place. And then I spotted the queen . . .

When you inspect the bees it's easiest if you draw one of the frames out, from the end closest to you, and then put it to one side. Then, because there are only ten frames instead of eleven in the same space, there's more room to look at the others, and you can slide the remaining frames forward and backward as required. When you put the last frame back, there's always a danger that one or two bees might get crushed, especially if you're wearing leather gloves, as your big fat leather-bound fingers get in the way and may hide any scurrying workers from view at the last second. To lose a worker or two is always a pity, but seeing as how there are tens of thousands of them in the hive during the summer, it's not exactly a tragedy.

The one you do have to keep an eye out for is the queen. As mother of the entire colony, she is rather important. Beekeepers pride themselves on their queens. They are precious, living jewels. Not only do they determine the nature and temperament of their offspring, but they have an aristocratic beauty. If they were human, they'd be tall with fine bones, long necks, exquisite figures and excellent deportment. They are not easily replaced. In other words, you look after them.

Although the queen stands out from the rest of the crowd through her size and shape, beekeepers help her to do this by marking her on the thorax; that's a round knobbly bit on her back, above where her waistline should be. Good beekeepers use a different colour of marker every year so that they can see instantly how old the queen is, and whether she will need to be replaced soon. Most of the rest of us use Tipp-Ex, the white correcting fluid beloved of typists. But Robin chose something rather more exotic: nail varnish, in an alluring shade of mauve. So she was hard to miss.

Generally, you don't have to worry too much about the queen once you've spotted her during the inspection, because she's quite shy. She'll beetle off down into the dark if she feels threatened, which is the best place for her when you're putting that last frame back in the hive. Because you wouldn't want her running round on the tops of the frames and getting in the way, would you?

Trouble is, later on in the summer, things get a bit sticky. To get ready for autumn, the bees start sealing things up with propolis, the resin that they collect from trees. It's their version of a draft excluder. It's very useful for the bees, but it's a bit of a pain for the keeper, as things don't glide together as easily as they did. Sometimes you have to give them a bit of a nudge. Or a shunt.

The thing is, I just didn't see her. I mean, I wasn't expecting to see her there anyway. Queens don't run around on the tops of the frames, do they? And as I gave the last frame that extra shove to put it in place, there was a flash of mauve. Under the frame. Robin was there with me, but looking at something else. I paused, hoping that I hadn't seen what I thought I'd seen. But when I lifted the frame, I saw her. Squashed. And rather dead.

I still feel tremulous about it now. There I was with the man who had taken me under his wing, teaching me every-thing he knew, sharing his bees and his expertise and his honey, and how had I repaid him? By killing one of his most precious belongings.

I stood, and sweated, and ran the following conversation with myself. In my head, of course.

Good Me: Oh, you idiot, you've gone and killed the queen.
Bad Me: No I haven't.
GM: Yes, you have. Look at it, it's dead. You and your big fat fingers went and squashed it.
BM: Maybe she's not dead. Maybe she's just lying down for a moment.
GM: In two halves? You're going to have to tell Robin.

BM: He hasn't noticed. We could just put the roof on and he'll never know.

GM gives BM a stern look.

BM: Okay, okay. I'll tell him.

Back in the real world. 'Er, Robin? I, er, I think I may have squashed the queen.'

'Where?'

'Here.' I lifted the offending frame to reveal Her Majesty, deceased.

Even now I can still remember the sound Robin made: an 'Oh' that was half sigh and half groan. It was the perfect expression for the moment. Surprise, regret and displeasure all rolled into one. He was far too polite and considerate to be angry with me. And he knew it had been an accident. But in turn, I knew how he felt. And it wasn't very good.

In time we got the queen replaced. Robin still let me help him with the bees. And I took extra care to make sure that when I put that last frame back, the queen was not in the way.

I wonder if it's partly redemptive to recall that what I have done unto others, I have also done to myself. I mentioned the marker, didn't I? Well, marking the queen is one of the trickier skills required of the novice beekeeper. And if you're a bit clumsy, as I am, it's a real challenge. First, we have to find her, and as she is at this stage unmarked, that can take some time. It's a bit like one of those *Where's Wally?* books, only in one colour.

Once you've found her, you have to hold the frame that she's on in one hand, while you reach for the catcher and the marker with the other. And when these are both tucked away in the bottom of your beekeeper's bag, this is nigh-on impossible. So you have to put the queen back, while you find the catcher and the marker, that little bottle of Tipp-Ex, unscrew the top, put it safely to one side, and start to look for the queen once more.

With a bit of luck you'll find her again within half an hour, though the bees will be a mite annoyed as you disturb them for a second time. At last, again, you have her, and again you lift out the frame that she's on, trying to keep her in view as you reach for the marker and the queen catcher. To be absolutely accurate, it's not a catcher but a 'Press-In Cage': a little round net the size of a fifty pence piece, framed by a circle of very sharp, vicious-looking spikes. The spikes are there to press into the wax comb. So you find the queen, place the cage over her, and press the spikes into the comb, which keeps her motionless while you mark her. It's quite effective, as you have to hold the queen still for a minute while you wait for the marker paint to dry. If she goes off too quickly, the workers will smell the paint on her. The odour of the paint may be stronger than her own scent, which could mean that they don't recognise her as their queen any more, and may attack or even kill her.

It's a delicate but necessary task, and one that the Bad Beekeeper approaches with reluctance. For a start, he knows how sharp those spikes can be, as he will already have punctured himself once or twice getting the catcher out of the bag. Anxious not to harm the queen, he will also hesitate, and hold the catcher too far from the frame while taking aim, and possibly miss the target altogether.

And yes, I've done these things, and worse. The first time I marked a queen, I was too tentative. I didn't press the catcher in hard enough on the comb, so she was able to move around a bit. In my haste to let her go I was a bit slapdash with the correcting fluid, and by the time I'd finished she was not so much marked as whitewashed.

Marking a second queen in another hive, I was determined to be firmer. I didn't like to see the queen struggle to be free. On the other hand, there was a job to be done. So I found her, held the frame securely in my left hand, picked up the catcher in my right, took aim – and plunged.

It was a perfect shot. Her Majesty was caught fair and square. One of the needles had gone right through the middle of her back. When I lifted the cage off, the spikes somehow came out cleanly and she was free to move around, which she did. For a couple of seconds. Not so bad, I thought, maybe she's all right after all. And then, as if remembering that she'd been skewered, she folded and died.

What were the chances? If I'd tried to do that deliberately, it would have taken a hundred attempts. But the gods of beekeeping were determined to teach me a lesson.

This would never happen to the good beekeeper, as he

would have read up properly on the subject first, and learned that the best thing to do is to practise by marking drones. They're big and easy to catch, and it doesn't matter too much if you kill one by accident, because there are hundreds more.

It's the lack of knowledge that can really do for the Bad Beekeeper. I learned that only too well from my third regal disaster. Yes, I've managed to send three queens to their doom. This last one was a more subtle failure, but the end result was the same. In the neighbours' garden, I'd opened up a hive that contained several queen cells which, as it turned out, were on the verge of hatching. One of them had already been born. Once I saw this, I should have shut up shop and left the bees to their own devices. For what usually happens is that the first queen out goes round to all the other queen cells and kills them one by one. It's a brutal business, but if it didn't take place there would be more than one queen in the hive, and heaven knows what sort of anarchy that would lead to.

Well, I was just about to do this, but then I saw something wriggling out of one of the other queen cells. Another queen was being born, right before my eyes. This was amazing. When you're inspecting the hives you can often see a worker being born; when two thousand are hatching out every day, it's hard not to spot them. But to see a queen emerging into the world is a rarity.

The joy of the moment was quickly tempered by the flush of panic. Oh, what to do?

I couldn't leave her in the hive, as she and the other queen would fight to the death. I couldn't just set her free as she would have no home to go to, and would probably die quite quickly out in the open. I felt responsible, as one would with any newborn. As I held her now, cupped in my gloved hands, she was, to my mind, just a baby. She needed to be tended to straight away. Time was of the essence.

Then I remembered that I had another colony without a queen in our own garden a stone's throw away. Perhaps they would accept her, and I would kill two birds with one stone. It was the simplest of manoeuvres. All I had to do was carry her over to the hive, hold her by the entrance and offer her up. All she had to do was walk in, which she duly did.

And I never saw her again.

The problem must have been with her scent. The bees inside would have detected this strange-smelling intruder waltzing into their home and, royalty or no royalty, would have done for her. What I should have done was to buy her some time: put her in an open matchbox covered with soft loo roll, and place it between the frames. By the time the bees got to her, they would have made their various introductions and taken her in quite peacefully. Oh dear.

To paraphrase the *Book of Common Prayer*: those are the things I've done that I ought not to have done. But I have also left undone those things which I ought to have done. Quite a lot of things, actually.

I should have kept notes. Experienced beekeepers tell you to do that right from the start, so that you know just what's happening in the hive, what you did last time, the condition and number of the bees, whether the queen is laying or not, how much food they have, etc., etc. Just a couple of lines each time you go to see them, so that you can remember what's going on. Otherwise you turn up a week later, sometimes two, and you have to rack your brain to recall why you stuck that drawing pin in frame number four. And how long ago was the new queen born in this hive – or was it the one next door? There's an awful lot to remember, especially if you've got several hives.

Naturally, the Bad Beekeeper doesn't keep notes. That would be far too efficient, and make life less complicated. No, he prefers to spend time at the apiary puzzling things out, especially in the spring when he has to guess which bees he had moved from where the previous season. It's so much more fun that way, and it wastes so much time.

Then there's the shed. Like many men, I only have to think of the concept of a shed and my heart lifts. It must be something about our caveman mentality – a nice dark place where we can seek refuge from the world. So when the time came to fork out a not inconsiderable amount of money on one, I didn't mind too much. I had too much equipment to store in the house, and there had been complaints. Personally,

I thought the hive parts standing in the hallway added a nice rustic touch to the house, but as usual on matters of domestic design, I was quite correctly overruled.

So the shed arrived; a lovely green wooden hideaway at the bottom of the garden. When it was still empty, I basked in its cavernous tidiness. This would herald a new dawn of apiarian (yes, there is such a word) orderliness. Things would be clean and neat, and I'd always be able to find just what I wanted in a trice, whenever I needed it. I had the same sort of feeling that some women get when they think of a pile of freshly completed ironing.

But I'm not a Bad Beekeeper for nothing. It wasn't long before stuff got piled in there. Lots of stuff, of indistinct purpose and unknown provenance. Job lots I'd picked up at beekeeping auctions that I thought I might need one day, once I'd worked out what they were actually for. Like that marvellous wax melter which turned out to be a retired surgical steriliser. An ottoman from my schooldays went in there, to store frame parts and wax foundation. There was a lovely wooden box I painted blue and filled with empty jars. A couple of old bookshelves lined another wall, for stacking with heaven knows what. Add to this the extractor, and the beesuits hanging from the rafters to keep them away from the mice, and before you know it the place was pretty full.

And that's before we chuck in the unmentionables.

If there's one thing you're told you really shouldn't do when you start beekeeping, it's storing old frames. Especially other people's old frames. It's all about hygiene. The bees, and later the wasps, will remove all the honey from any frames left lying around. But the wax comb left behind can easily contain traces of disease long after the inhabitants have left. It's all in the spores. So keeping old bits and bobs from unidentified sources is very, very naughty.

Well, of course I have.

I couldn't help myself really. I think I get the habit of collecting useless bargains from my mother. As a child, I remember we would trawl around auction houses and places like old schools that had closed down, looking for things which 'might be useful one day', like wardrobes or chests of drawers. For years in my parents' garage, we would trip over a pair of wooden beams taken from an old gym. They could serve no earthly purpose, but we probably got them for nothing, and so there they lay.

That's the label that should be hanging on the boxes of frames I've had in the shed for years. They had been in some old abandoned hives that I'd taken over on a farm, and it just seemed like a terrible waste to get rid of them. So they've been stored in cardboard boxes in a corner for several years, covered in cobwebs and old plastic bags containing lumps of wax, and unidentified bits of tin. The daftest thing is that I can't even use them. They're the wrong type of frame for my hives.

The shed is the beekeeping equivalent of Tracey Emin's unmade bed. A recent inventory of the contents revealed the following:

A box of cut comb containers, bought five years ago and as yet unused. A giant ziplock bag. A small box of fumidil – an antibiotic years past its sell-by date. Various samples of nutrient powders and gels, picked up free at a beekeepers' convention. A container of white crystals. A bottle of formic acid (don't remember where that came from). A plastic jar with some unidentified white stuff in it, origin unknown. A dried-up sponge wipe. An orange strap. A cut comb cutter. Boxes of tacks and hive staples. A jar of mysterious-looking syrup. A small container of 'beeswax balms' (I've got no idea, either). Lids, various. A brush. The bottle of Bee Quick I thought I'd bought six months ago but couldn't find. More ziplock bags. Five empty coffee tins. Some newspaper. A plastic bag with some very mouldy old wax inside. Dozens of plastic frame spacers. A carved wooden sign that says 'Billy's Bee House'. Two old golden syrup cans. And a primitive representation of a beehive that I attempted to paint as a guest on a children's art programme, which is so bad it's not allowed in the house.

And that's just in the cupboard to the left of the door.

I have to say, I'm not proud of this. Of all the bad things I've done as a Bad Beekeeper, the long process of stacking all that stuff down there is probably the worst. For the fact that I could let such a collection of useless crap build up over such a length of time probably represents something fundamental about the rest of my life. I'm a slob. Cleanliness is next to someone else's house, not mine. Left to my own devices, and without the guiding hand of a benign spousal authority, I would lapse far too easily into slatternry and slovenliness. Perhaps subconsciously, that's why I love the bees so much; they're so neat and tidy. A model to which one can aspire.

Let's face it though, most of us at some point have dark little corners in parts of our lives. Mine just happens to be in a green wooden shed at the bottom of the garden.

So you see, Mildred, I really can't marry you now, can I?

Chapter Twelve

REDEMPTION

*Where we discover that good things can
come of bad beekeeping.*

Fortunately, I have been able to make up a bit for the bad stuff. The road to redemption started a few years ago with a trip to the Beekeepers' Convention at Stoneleigh in Warwickshire. Run by the British Beekeepers' Association, this is the Boat Show of the beekeeping world. They come from all over the UK to snap up the bargains offered by the big retailers, see the latest gadgets and equipment on the market, attend lectures by experts, and swap notes with their fellows. It's a pilgrimage that all beekeepers make sooner or later. Some go every year, others just when they can afford it.

So there I was in the throng, with my plastic bag full of free samples that I would later store in the shed and forget about for years afterwards, wondering if it really was worth investing in a German-built honey warmer with thermostatic controls, stainless steel lid and . . . when I came across a stand that caught my eye. Bees for Development. Not the catchiest name in the world, to be honest. Not exactly Oxfam. Or Children in Need. Still, it raises a question. Whose development? So, as happens so often at the convention, I stopped to chat with the two women who'd founded the organisation some years before.

Nicola Bradbear has spent her life working with bees, and getting bees to work for the poorest people of the world. She travels regularly with the UN to assess the needs of beekeepers in far-flung and often troubled corners of the globe – Iraq, Afghanistan and Chechnya to name but three.

In 1993, Nicola founded Bees for Development with Helen Jackson, a biologist. They had a simple goal: to assist people living in poor and remote areas, and to raise awareness about the value of beekeeping to alleviate poverty. This latter aim may sound a bit tortuous, but it makes sense. Nicola's theory is this: if you live in a remote village and you're doing comparatively well, you might own a cow. If you can't afford that, you might own a goat. And if you can't pay for one of those, what you can afford – what anyone can afford – are bees.

The aim is not to give people beehives or equipment. European beehives are not suitable in Africa and would have to be replaced in time anyway, and there can be complications with maintenance and ownership, and debt. Bees for Development doesn't give anything that you can lay your hands on, apart from its regular journal. What it does provide is knowledge and expertise: a free information and advice service to beekeepers in developing countries. So it distributes technical information and know-how, keeps beekeepers up to date about events and training opportunities, carries out research and lobbies for changes in government policies to make trading easier. From the simplest things, such as how to make a basic hive for African bees and the best way to take your product to market, to international tariffs on the import of honey into Europe.

Nicola has found that the problem for beekeepers in many areas is not one of equipment or producing honey, but how to sell it. Developing market connections sounds very dry, but for many people it's the key to making progress.

So when I decided to run a marathon wearing a beekeeping suit, they were the obvious beneficiaries.

Perhaps I should back up on that just a bit. I was in the habit, a few years ago, of running the occasional marathon. I started with the first London race in 1981. I'd covered the press conference when the plans for the event were revealed, and got this crazy notion to try to do it myself. In those days, there were hardly any marathons. New York, Boston, maybe one or two others, and that was it. Shops selling 'proper' running shoes and training gear were few and far between. I remember having to trek right across London to buy my first pair of trainers. The first race started at the gate of Greenwich Park, with what was thought to be a huge field of eight thousand runners. Like all first timers, I was just anxious to finish. And when I crossed the line in four hours and twenty-four minutes, I reckoned that was the end of my marathon running career.

Years passed – eighteen to be precise – during which I went to live in America twice, got married, had three children, travelled the world and pretty much forgot about marathons. Until, on our return from four years in Washington, a neighbour somehow convinced me that not only could I run twenty-six miles and 385 yards, but that I would *want* to as well. I'm not sure whether time heals all wounds, but it does dull the memory of pain; so I agreed, and found myself on the starting line for the London Marathon in 1999. And 2000. And 2001. I'd probably have kept going, fixated with the idea of achieving a new personal best every year (and failing), until a colleague said quite simply, 'You know, you don't *have* to do it.' The scales fell from my eyes, and I hung up my running shoes. Actually I put them under the stairs, but you know what I mean.

So I'm not quite sure why I decided to do it again in 2005. But there I was, pounding the winter pavements for eight miles in the dark *before breakfast*, and thinking that it wasn't such a bad idea. Then injury struck. There was something wrong with my right foot, which put paid to training for several weeks and any thoughts of a new personal best.

I needed some new motivation. I also needed an excuse for running what would be my slowest marathon to date. And then inspiration hit me, like an angry bee smacking into a veil. Of all the tens of thousands of fancy dress runners in the event, among the Elvises, and pantomime horses, and fairies, and daffodils, and Wombles, I'd never seen anyone run the London Marathon in a beekeeper's suit. What a jolly jape that would be. And a good fundraiser for Bees for Development.

The good people at Thorne's, the beekeeping suppliers, knocked up not one but two special lightweight suits for me to run in, with gloves and veil – the whole deal. I was a vision in white, but decided to do without the smoker, for I'd be doing plenty of puffing on my own. Normal runners wear the skimpiest of clothing for their races, with good reason. The body temperature can climb way above normal, so the best uniform is a singlet, short shorts, lightweight socks and trainers. Long trousers and a long-sleeved jacket on top of a vest, plus leather gloves and a veil encapsulating the whole head don't make for the best ventilation.

I tried a couple of long-range practice runs over the fields and streets of Buckinghamshire. One route took me through the middle of Old Amersham. You'd think, wouldn't you, that a man pounding along the pavement in a full beekeeper's suit and veil might raise an eyebrow. And from a distance, one or two people stopped and pointed, as if to say, 'Wow, there's a nutter running in a beekeeping suit.' Interestingly, close-up there was no reaction. I ran right past one couple who must have seen me coming and didn't bat an eyelid, which meant that either (a) I had become invisible, or (b) they had spotted the nutter in a beesuit but didn't want to provoke any reaction from someone who was already clearly disturbed.

The day of the big race was warm and sunny. And it wasn't long before I felt the need to divest, just a little. After a mile the gloves came off, literally, but I felt I had to carry them for honour's sake. Plus they were a good pair of leather gloves and they'd come in handy another time. But after two miles, what with the general jiggling and joggling going on, I dropped one, and nearly caused a disaster.

In any race, the participants are clustered pretty close together at the start. But with 40,000 runners, the London Marathon is so big that it needs two separate starting routes, which only come together after three miles. It is a very tightly packed bunch of people. And in the early stages, with the adrenalin pumping, people are moving at a fair old pace. It can be hazardous. If someone trips or stumbles in front of you, there's very little time or space to manoeuvre round them. And occasionally, there are pile-ups. The stakes were heightened by the fact that I had managed to get into the celebrity start, right at the front. So there I was, a forty-nine-year-old man in a beekeeping suit, being overtaken at speed by some very fit, fast and eager marathon runners.

Not, then, the best time to drop one of my gloves. Or rather, to stop and attempt to pick it up. It's instinctive, though, isn't it? You drop something, you pick it up. Had they been ordinary gloves, I might have simply sighed and carried on. But these were special. Lightweight leather gauntlets which would later come in handy at the hive.

To be honest, I didn't even think about it. Glove drop, feet stop. Bend down. Hear screams. Look around. And see a human stampede about to fall on top of me. Somehow they managed to part like waves around me in the two seconds it took for me to stoop, pick up the glove, and carry on. For the next minute all I could say was, 'Sorry, sorry, sorry,' to everyone who now passed me, though the people I'd nearly tripped up were long gone. Still, perhaps it was a talking point:

'Did you see the weirdo in the beekeeping suit who kept apologising to everyone?'

One of the things you notice early in a running career is the need to expectorate. When you get warmed up, the mouth creates a lot of saliva, which then has to go somewhere. I've always found that outwards is the best place. Otherwise, it just hangs sloshing around as you jog along, getting in the way. Spitting isn't pleasant, I know. But when you're out with 39,999 other runners, you can reckon that at any second at least a hundred of them are doing the same thing, so you're in good company.

Shortly after the narrowly-averted-glove-dropping disaster,

I felt the need to let one go. So my tongue gathered the requisite fluid, my lips tightened and . . . *splat!* A big, fat, juicy wodge of saliva exploded in front of my face. I'd forgotten to take my veil off.

After six miles, I felt I was beginning to overheat a bit. They don't talk about a 'head' of steam for nothing, and inside the sealed confines of the beesuit the temperature was indeed building up. I was worried that people along the route wouldn't recognise the purpose of my outfit, even though it had the words 'Bees for Development' embroidered in large letters on the front. So I made a deal with myself to take the veil off during the quieter periods, and wear it through the busier sections where the crowds were gathered, like at Canary Wharf.

Deal-making was a big part of marathon running for me; negotiating with oneself, principally over when to slow down and take a breather. No walking before Tower Bridge was the cast-iron rule. After the halfway point, though, the talks could begin. Although 'negotiating' is a rather grand title for what was really whining from the inner child. Anyone who has ever taken a toddler on a walk will know the routine:

'I'm tired . . . I'm really tired . . . *Carry* me . . .'

So you come to an arrangement. At the end of mile fourteen, you can walk for a minute. That helps a bit. By the end of mile sixteen, though, walking doesn't do it any more. You want to stop. A mile later, you want to sit down. By the time you're at twenty, you just want to lie down. And the road is hard, and your feet hurt, and your body, sealed inside its protective suit, is hot and clammy, and your stomach is sloshing full with all the water you've taken on board, and yet somehow you're still thirsty, and now you feel slightly sick from the energy drink you had at the last fuel stop. And you wonder why you decided to do this in the first place, and had you forgotten how ghastly it was going to be, and you're kidding yourself that the last six miles aren't that long really.

But the crowd does lift you, especially when people call your name. They could hardly miss it, as the letters BILL were stitched loud and large across my chest. Passing Embankment, I could hear one particular voice ring out. I turned back to see Sophie Raworth, one of my old *Breakfast* sofa partners, on the pavement. It was hugely cheering to see a friend on the sidelines, though kissing her through the beekeeper's veil was rather complicated. Not so much a peck on the cheek as an abrasion. Never mind, the spirits were lifted. Two miles to go and the crowds were thicker now, and louder. Soon we were all picking up the pace for the finish, charging down Birdcage Walk towards Buckingham Palace, around the corner and the last agonising sprint up-hill to the finish line, and it's over. The veil came off for the last time and I made that same plea to myself to never, ever, do that again. Afterwards I noticed a red stain on my chest. It looked as if my heart had been bleeding for the cause. As it happened, it was a simple case of chafing; my nipple had bled through two layers of clothing, and I hadn't felt a thing.

It's the honey, and hence the bees themselves, that have done the most on my path to redemption.

It all started with a talk I gave about beekeeping in a little village hall close to home, on behalf of a local hospice charity. As you can imagine, given how much I know about the science of apiculture, it was a short talk. So to pad it out a bit, I offered a jar of my honey for auction. To my amazement, the bidding got as high as £30. I was almost embarrassed to be charging so much for my own product, so I gave the highest bidder two jars for his trouble.

That got me thinking that this might be a useful way to raise money for charity. In the past, I'd sold jars of honey to people at work for good causes. But the auction idea was a brilliant way of getting a lot of money in a short space of time. I had a chance to try it out a little later at a rather swanky event in London. Every year I'm invited to host a ball on behalf of a charity called African Revival, which is attended mostly by people in the insurance industry. And every year, as with most charity balls, they hold an auction. Footballers' shirts, weeks in country cottages, the chance to drive a sports car for a day, that sort of thing. Against that lot, I didn't think a simple jar of honey would get much. But I suppose it did have some rarity value, especially given the amount of honey I'm actually able to produce every year. Jokingly I suggested that they might want to break the record of £30, and in no time they did. One hundred pounds was what we got. I was truly impressed.

So now I had, to coin a phrase, a ball rolling. A couple of months later I was asked to host another charity event – a Gala Ball organised by my *Strictly Come Dancing* partner, Karen Hardy. I could hardly refuse, could I? And again there was an auction, and again I took along a jar of honey. The audience here weren't the high rollers of the insurance world, so I didn't really expect them to rise to the challenge of setting a new record. Silly me. Almost right out of the traps the bidding crashed through £110 and rose steadily thereafter. Two hundred. Two hundred and fifty. Two seven five. Three hundred. This was getting exciting. Finally a big jump to £400. Four hundred quid for a single jar of honey. I, along with everyone else in the ballroom, was quite simply gobsmacked.

The next year the insurance brokers had upped their game for the African Revival ball. We were dining and dancing at the Guildhall in the City of London, a grand cathedral-like chamber, rather like the assembly hall at Hogwarts School.

'The record,' I announced with pride, 'stands at £400. I know you can beat it.'

The initial ascent didn't take long. We were past the target within a couple of minutes, and climbing with ease. Five hundred, six hundred, seven hundred and fifty. This was amazing. The bidding went up, but the number of bidders was gradually whittled down. As we slowed down in the seven hundreds, I wondered if we could make the magic four-figure mark. I needn't have worried. Once we got to £950, the summit was too tempting to avoid. And there we stopped. One thousand pounds. A grand for a jar of my bees' honey. I felt almost bewildered at such an amount, but also very proud of my girls.

It was slightly deflating, then, when the next lot sold for the same amount. A green raffia lizard, almost a joke item. It went to a Russian blonde. Afterwards I asked her why she'd bid so much for something that she was clearly not very keen on. 'I have to spend my money on something,' she said, and turned back to her friends.

It was September 2007. A sign of the times.

A year later, it was a very different world. Lehman Brothers bank had collapsed in America, and Northern Rock had nearly done the same in Britain. The financial markets were in meltdown; disaster loomed at every turn. I took another jar of honey to another African Revival ball, thinking I'd be lucky to get half the amount we'd raised twelve months earlier. But I'd reckoned without the stubborn loyalty of two particular individuals. One had a son who'd spent some time on a project in Uganda that summer, and had been involved in a beekeeping scheme out there. The other was the £1,000 man, who'd bought my jar the previous year.

It had been decided to put two jars together as one lot; one from the Ugandan project alongside my own. I wasn't sure how well that would work. So naturally, it turned out to be a stroke of genius. Separately, in those economic conditions, they might not have raised much. But together they sparked an almost lunatic competition between the two men: one of them driven by pride for what his son had achieved on the African project, the other determined to keep his 100 per cent record when it came to bidding for Bill T's honey.

I can't remember where we started, but I knew we were heading somewhere fairly spectacular when the bidding raced through £500, and sped on through the £1,000 mark. We went on in leaps and bounds. Sometimes we rose by a hundred, sometimes by half a grand as the two combatants (the others dropped out quite quickly) sought to knock each other out. As they were sitting in quite separate sections of the room, it took a little while for me to walk from one to the other to confirm their bid, which added to the tension and gave them more time to consider their next move. Two thousand came and went. Two and a half. I couldn't really believe this was happening. Three thousand. All I could do was manage the bids, as they went up in increments of at least £500 a time. Three thousand five hundred. Four thousand. This was spellbinding. Normally at one of these events, there's always a bit of a buzz as those uninvolved chat to their friends. It can make life difficult for the auctioneer. But now I had no such problems, as all in the room were transfixed by the duel. Four thousand five hundred. As I strode from one table to another looking for the next counter-offer, I wondered if anyone was really going to pay this kind of money. Maybe it would all turn out to be a hoax.

But it wasn't. Instead, something even more extraordinary was about to happen. As we looked for a bid of five thousand pounds, the man who'd bought my honey before beckoned me over. 'Look,' he said, 'I only want your honey. Let's just split it at five.'

That seemed fine to me. Both bidders would get what they wanted for £2,500 apiece, and the charity would take £5,000 in all.

But that wasn't what he meant. He was suggesting that they both pay the full amount, £5,000 each, and split the auction lot down the middle. He'd have my honey, and the proud dad would have the Ugandan jar. In other words, the bid was being doubled to no less than £10,000.

And so it was that the nectar drawn from the apple blossom on a farm in Buckinghamshire was turned into honey by the bees in the hive, and extracted and poured into a jar in the kitchen of my house, and ended up helping to pay for a school in Africa. Funny old world.

There was one other auction incident that stood out for me that year, not so much for its financial as its comic value. It's a long story, so you might want to sit down. This time it involved a jar of honey (of course) and a certain Chris Tarrant:

the long-serving morning radio DJ, host of *Who Wants to Be a Millionaire?*, general mischief-maker and Lord's Taverner.

Whenever our paths had crossed in recent years, we'd usually engaged in some good-natured verbal jousting. We'd met at Buckingham Palace, at a garden party to celebrate fifty years of the Duke of Edinburgh Awards. Tarrant and I, and a host of other celebs and media people, had been involved with the Award programme on and off, hosting ceremonies at St James's Palace. This involves handing out certificates and doing the warm-up act while waiting for Prince Philip, and more recently Prince Edward, to come round and talk to the worthy awardees.

As a reward for our good behaviour, we were invited to the Palace along with thousands of other people, to see, and in our case to meet, a senior royal. While others were to shake hands with Prince Philip or Prince Charles, I was particularly lucky, being placed in the central avenue down which would come the Queen herself. We were penned into groups of four to meet her. In my group: Tim Smit (CBE), founder of the Eden Project; Alan Whicker (CBE), veteran broadcaster; Chris Tarrant (OBE), and me (No BE).

Our four, however, had become three, as Alan Whicker had gone AWOL, and no one could find him. This was good news, as it meant more precious royal seconds would be devoted to the rest of us. As we waited and watched Her Majesty make her way down towards us, Tarrant and I laid a wager.

'Bet she talks to me for longer than she speaks to you,' he said.

'You're on,' I replied. It's probably not really the done thing to lay bets in the gardens of a royal palace as the monarch approaches, but I blame it on Tarrant's compulsion for mischief. If he hadn't been there, I'd have been perfectly well behaved.

As befits a man who's done something decent with his life, like setting up the world's largest greenhouse, Tim Smit was introduced first. 'May I introduce Tim Smit, Ma'am? He established the Eden Project in Cornwall.' A noble venture, a fine man, clearly worth of Her Majesty's interest and hence a good two minutes of question and answer.

Next, CT. 'This is Mr Tarrant, Ma'am, who has a quiz show on the television.'

The hand was shaken, a greeting was murmured, and that was it. No conversation. Tarrant was out for a duck. He'd missed the crucial penalty. He'd been forced to take the early bath. He was already on his way back to the pavilion.

All I needed now was a sentence, a phrase, a raised eyebrow even, and I was in.

'And this, Ma'am, is Mr Turnbull. He reads the news and has a bit of a penchant for ballroom dancing.' I felt like Wayne Rooney on the edge of the box, taking delivery of the perfect pass, poised to shoot.

'Dancing?' Her Majesty enquired. Rounding the goalkeeper, the striker has only the empty net in front of him.

'Oh yes, Ma'am.'

And I was off.

'Itookpartinacompetitioncalled*StrictlyComeDancing*acou pleofyearsagoandwasn'tverygoodbutgotluckyasIpickedupanin juryandpeoplevotedformetostayinthecompetitionacoupleof weekslongerthanIshouldhave.'

The Queen smiled patiently. This probably happens to her all the time. It wasn't much of a dialogue, but it had been one. And it meant that I had won.

No money was exchanged. It was worth much more, because it meant that whenever I bumped into Tarrant after that day, as I did from time to time, I could relate the tale of the Buckingham Palace encounter to all and sundry and bask in my victory once again.

Until we came to Sark, that delightful little island in the Channel where there are few people and even fewer cars. The purpose of our trip was to raise money for the Lord's Taverners, who for some reason had invited me, a non-cricketer, to take part in their festivities alongside the cartoonist Bill Tidy, Nicholas Parsons, Robert Powell and the Tarrant himself. It was a lot of fun. We were a merry throng, especially in the evening when we gathered for a dinner and yes, an auction.

One of the best prizes on offer was a week's stay at a winery in South Africa. When the bidding started, I had no intention of getting involved. But Tarrant was getting stuck in, and was pretty much on his own. He was going to get a wonderful trip to a gorgeous place for a song. Outrage bubbled up inside me, along with a couple of glasses of wine. I turned to the Supportive Spouse and said, 'He can't possibly have it for that little. Shall I bid?'

Supportive Spouse, similarly fuelled, nodded her assent.

I had no real intention of claiming the prize. I just wanted to make Tarrant work for it. So why, then, did I blurt out an offer that was £500 higher than the previous bid?

Tarrant and I stood up, facing each other down across the marquee like gunfighters. I tried to bristle manfully, muttering, 'Come on, Big Man, show us what you've got.'

Inside though, I was hoping to communicate telepathically with my good friend Chris. *Don't worry, mate, just kidding. Make another bid and I'll sit down.*

This, though, was where I was going to pay for Buckingham Palace.

Three hundred pairs of eyes were trained on Tarrant as he summoned his response.

Then slowly and surely, he raised his arm like an umpire, and pointed it at me.

The room went quiet.

'It's yours.'

On the face of it, I'd won. But we both knew I hadn't really. And as the crowd cheered, the squeaky little voice of my inner child whined. 'But I don't want to go to South Africa.' No matter, a week for two in the Cape would be a nice break for Mrs T and me; just the two of us. Except that it turned out that what was on offer was not a week for two, but for the whole family. And with air fares, the trip would end up costing five times the amount of my winning bid.

Round two to Tarrant, then. But there was to be more the following day. Another meal, another auction. And this time I had my secret weapon: yes, a jar of the special stuff. I'd spent the previous couple of days priming one or two other members of the trip to open the bidding, so that I wouldn't be totally embarrassed if no one wanted to buy it.

They handed me the gavel for my item and off we went, rising up through the hundreds with grace more than speed. I could sense the steam was beginning to run out when we laboured through the five hundred mark, chugging up to six, and sputtering into seven, before grinding to a halt. At which point I drew my trump card.

'So, where is Chris Tarrant?' I boomed, like the headmaster in assembly, calling out the name of the boy who broke the chapel windows.

Heads turned, and a reluctant hand went up.

'Can I take it that's your bid, Chris? For eight hundred pounds?'

I assumed from the grizzling and muttering that was emanating from his visage that he was indeed confirming his generosity; and the jar, gloriously, was his. Round three to the Bad Beekeeper.

Fate, though, was not done yet. I had to leave that afternoon, to get home in time for work the next morning. The rest of the party stayed on to complete their cricket match, and head back to London the following day. After the ferry from Sark, Tarrant duly checked in for his flight from Guernsey and proceeded through security, where he ran into a hitch. As he put his bag through the X-ray machine, a jar popped up on the screen. A jar containing honey. Liquid honey. More than one hundred millilitres of it; which meant, according to the anti-terrorism regulations, that it was big enough and wet enough to pose a deadly threat.

Tarrant on his own can look quite combustible at the best of times, so it was not an entirely unreasonable call. Sadly, there were no independent witnesses to the incident. But I can imagine there was much fulminating, remonstrating, and general bad humour; all to no avail. If he wanted to fly, Chris would have to give it up. The mighty Tarrant flew home. And his £800 jar of honey stayed on the island of Guernsey.

The best part of it was that he doesn't even like the taste of honey.

In case you should feel sorry for him, there is a heart-warming postscript. A few weeks later at another Lord's Taverners event, there was yet another auction, and yet another jar of honey for sale. Tarrant resolutely refused to bid. But a kind fellow in the audience snapped it up for a mere £300 – a bargain – and handed it straight over to CT.

Makes your heart melt, doesn't it?

SEPTEMBER
INTERLUDE

It is late September. The sun is shining brightly out of a clear blue sky. On the ground a dusting of dry brown leaves, as the first trees shed for the autumn. When the light breeze strikes, there is the hint of a chill in the air. But it's still warm enough for shorts and T-shirts. And I am taking honey off. Two months late.

Around the hive I'm visiting there is almost frantic activity. A horde of bees gathered around the front and the side. This is a colony I put together with two queens during the summer, and have left on their own for weeks. I'm not sure what's going on. Are these clouds of bees trying to get in, in order to rob the place for the contents inside? Or are they just working?

I crack open the boxes, wedging the sharp chisel edge of my hive tool into the corners to break the seal of the propolis, and gradually it comes away, still sticky with the warmth of the Indian summer. Inside there is a fizzing, which subsides as I spray the bees coating the tops of the frames with sugar syrup. As the silver drops land on their backs, they quieten almost instantly and, crowded as they are, I can delve in amongst them, pulling out frame after frame to see if there's honey here worth taking.

Amazingly, given the crime that I am perpetrating, the bees are remarkably compliant. Occasionally one will stray onto my gloved finger and inspect my hand. But a quick flick of the wrist and they fly away.

The frames that are only partially filled with honey I will leave. You can feel almost instantly how much is there just by lifting one lug on either end. A satisfying weight on the tip of the finger indicates that there is booty here for the taking.

I lift the super off and put it to one side. The lighter frames, with bees attached, are put back. The heavier ones are shaken and brushed clear of their guardians and taken clean away. I can do this today without danger, as there are not too many bees on this part of the hive.

Somehow they don't seem to notice, or they don't seem

to mind. I have made away with my haul and yet am still at one, more or less, with the colony.

In the autumn sunshine, it is a golden moment. And I can pause to reflect that this, too, is what it's all about. It's not always like this, not by any means. But just once in a while, even the Bad Beekeeper gets away with it.

Chapter Thirteen

THE BAD
BALLROOM CLUB

In which the author takes a break from the ardours
of apiculture to try his hand at a televised terpsichorean
contest, and endures further suffering. His bees, though
neglected, enjoy the break.

One of the worst things you can do as an apiarist is to abandon
your bees. It's not as if they'll starve; in fact they might survive
for some time on their own. But the chances are that in this day
and age, the varroa mite or some wicked disease will put paid
to them before too long. There can be exceptions, though. My
most successful colony is descended from a hive I adopted from
the outer reaches of a large garden. The owner of the property
was a friend of a friend, who said the person tending to the bees
had not appeared for five years, and would I take them away?

For a moment, I wasn't sure there were any bees there anyway, as the whole area was covered in brambles. But if you looked hard enough you could just make out the contours of a couple of hives. And there, climbing vertically out of the undergrowth, was the occasional bee. It took a good hour to hack in there, but what I found was worth the work. A full, healthy colony that had just been doing its own thing for several years. In fact they'd done their own thing so successfully that they'd gummed all the frames in the brood box, making it impossible to inspect them. To check the bees to see if they've got enough food, to discover whether the queen is laying, and to make sure they're all healthy, you need to lift each frame up one by one and examine both sides. But the frames in this box were so well stuck down that any attempt to lift a frame simply snapped the lugs on either end, ruining the chances of any further inspection. I was able to give them some preventative treatment in the form of the thymol paste used to ward off varroa. But otherwise it looked as though they were there for good.

How had these bees managed to survive, when so many others might have died out? I simply had to deduce that they were, to use Gavin Pritchard's phrase, 'local bees'. And I've used them as the dominant strain to breed from for all my other hives. When I say 'breed', it sounds rather grand and scientific. In fact, it's nothing of the sort. It just means that when they make queen cells, they're the ones that I use to develop other colonies.

While they are my best and most productive bees, they nearly didn't make it. One year they became seriously bad-tempered after I'd taken the honey off, to the extent that you couldn't go near the hive without getting buzzed and followed by some over-attentive outriders. And when the farmer mentioned that they were bothering him as well, I knew something would have to give. Just as you don't bite the hand that feeds you, you shouldn't sting the farmer who's housing you either. They were now so vicious that inspections became a real battle. Along with all the other hives, they settled down for the winter. But I knew that come spring, drastic action was called for.

What I needed to do was to replace the queen, or simply remove her and let the bees make a new one. Old queens can make for bad-tempered colonies. A new one, with a slightly different set of genes, can usher in a better mood. Rather like a coronation in our own world, except that the bees don't know how to make bunting.

The problem was, though, that as this colony had been on its own for so long, and had in effect redesigned the layout of the brood box, I couldn't get in to find the old queen. So replacing her was impossible.

The only other option was a sentence of death. I couldn't face a whole season of bad-tempered bees. And I certainly couldn't have them attacking anyone else. So it was time to get the fuel canister out. The one thing that will kill a bee faster than anything else is the vapour from petrol. Well, squashing is effective of course, but it would take a while with tens of thousands of bees, and they wouldn't exactly be lining up for it. So what you do is get a small flat container, the size of a saucer, pour some petrol in it, and slide it into the entrance of the hive. I'm told it takes just a few seconds. But I'm also told that the sound of a full colony fanning desperately at the fumes and then, as one, giving up the ghost is not pleasant to hear.

So it was with a heavy heart that I approached the hive the following March, with my little plastic container and its deadly colourless contents in my hand. When I got to the front to administer the poison, I noticed that something was different. There were bees flying in and out of the hive, which was not unusual. But they weren't attacking me. Several months previously they would have been bouncing off my veil like pellets if I'd dared to stand in front of their entrance. But today, they didn't seem to mind. I put the petrol on the ground, taking a moment to have one last look inside. I lifted off the lid and they were all there, as meek as lambs. It was the most extraordinary conversion of temperament. Perhaps they'd replaced their queen late in the autumn, though I doubt they'd have had the time. Perhaps they'd just decided that bygones should be bygones, and that they would give me another chance. The petrol was taken away. The execution was cancelled, and a general pardon was bestowed.

Shortly afterwards I managed to get them all into a new brood box by driving them up with copious amounts of smoke. I can now inspect them properly, and make sure they have what they need. And since then we've been the best of friends. Well, sort of.

But as I was saying, one of the worst things you can do is abandon your bees. Sometimes, though, it just can't be helped.

It was the last day of August 2005; a Wednesday. I'd presented *Breakfast* that morning with a programme dominated by news of Hurricane Katrina, which had devastated the coast of Louisiana. I had in mind to take some honey off the bees that day; late as usual, but the weather was still fine, and the honey was still there. First I had to take a nap. It's something I do every day I'm working. If I don't lie down for a while, I'm a basket-case by teatime.

So there I was in the land of nod when, most unusually, my son came into the bedroom with a message. 'Dad, your editor's been on the phone. They want you to go to America to cover the hurricane. Now.' The honey was going to have to wait.

This was at 1 pm. At 4.30 the plane took off from Gatwick. Things do move fast in the world of news. So while I should have been processing honey from the gently shaded woods on the farm, where the dappled sunlight played on the apple trees, I found myself instead surveying the wreckage of the largest natural disaster in the history of the United States. Hundreds of miles of coastline in Mississippi and Louisiana had been ripped up by winds of 175 miles an hour. At the time, Katrina was the most powerful hurricane to hit the Gulf of Mexico, and it showed. We started our coverage in the resort town of Biloxi, where the flood waters had significantly rearranged the architecture on the sea front. I'll never forget the sight of two enormous riverboat casinos, mounted on vast concrete platforms, which had somehow been lifted off their moorings and floated two hundred metres inshore. We drove down a road where virtually every house had been damaged in some way. A huddle of remaining residents were consoling themselves with alcohol. When your whole world's been blown away, and there's no electricity, no running water, and no civil authority, some people reckon the only thing to do is have a party.

A couple of days later we managed to work our way into New Orleans, a city largely underwater and almost completely abandoned. To reach the centre we had to drive through a maze of routes, having to back up and start afresh each time we hit water. It took hours, but eventually we got to Canal Street, the heart of the city, which had been completely trashed by looters days earlier. If you wanted to see a vision of a post-apocalyptic society, this was it. What had been the party capital of America was now its biggest ghost town. You could walk along the six-lane motorway that encircled the city centre for as long as you liked in perfect safety, for there was not a single vehicle on it. Incredible.

A week later I was home, hoping to pick up where I'd left off, and complete the honey harvest for the year. But before I could get out to the hives again, there was an email waiting for me, one that was to change my life in the most remarkable way.

'Dear Bill,' it read. 'We are currently working on the new series of *Strictly Come Dancing* and we would like to see if you would be interested in taking part in the show as one of our celebrity dancers.'

Now I knew I was in the Twilight Zone. First, a cataclysmic hurricane, then the whirlwind of taking part in what was then the most popular show on TV. But it was no dream. My first impulse was to say no. I'd been called many things in my time, but 'celebrity dancer' was way off the scale of my professional spectrum. I was a news man, dignified, serious . . . ish. I had a reputation to protect. On the other hand, it would be a lot of fun. My children thought I was mad to be even thinking about turning it down.

'Dad!' they cried in unison. 'It's a no-brainer. You've *got* to do it!'

Still cautious, I decided to consult higher powers within the BBC news machine. If any one of the three people who constituted my bosses said no, I'd have a good reason to turn it down. Their judgement was unanimous. No objection. I might never get to present *Newsnight*, but that was hardly on the cards anyway, so I might as well go ahead.

It was the best professional decision I ever made. From

that moment I had no regrets. It was the craziest, funniest, most exhausting experience of my life. And I had no idea what I was letting myself in for.

I hadn't really watched the show much in the first two series. I'd seen Natasha Kaplinsky win the title with Brendan Cole in the opening year, and knew straight away that there was no way I could ever win. I just reckoned on turning up for a couple of practice sessions a week, bobbing along on a Saturday night, and maybe getting through to week three if I was lucky.

But then I had reckoned without the power of my dance partner, the formidable Karen Hardy. And that was my first mistake. She walked into my kitchen with a camera crew behind her, went through the pleasantries of our introduction, and got her diary out.

'So, when shall we start?' Karen asked. With the competition three weeks away, I thought we could leave it for a while. Another mistake. 'How are you fixed for Monday?'

And so began a rigorous schedule of rehearsals, three to four hours a day, five days a week. Ostensibly we were starting with the Cha-cha-cha, but in all honesty it felt more like boot camp. I was horribly unfit for dancing, even though I'd run a marathon only five months earlier. There were challenges on every level. Learning the steps was just a small part of it. The body had to be taught to move more quickly and in different directions than it had ever done before. I was told I needed to learn a new way of walking, which with the added inch of Cuban heels (for Latin dances) was difficult enough already. And this was just the physical side of it.

I should say at this point that Karen was delightful to work with. But she was firm. She'd had a baby the previous year, and was returning after five years of retirement. As a former international champion, someone who'd been at the top of her field, coming back was a big deal for her. And here she was, lumbered with a forty-nine-year-old journalist who hadn't danced ballroom since he was twelve.

'I've got to tell you, I'm very competitive,' she said in our first session. 'I don't do second place.' Well that was okay, because there were eleven other positions we could fill, from twelfth up to first.

I soon found that I was having to tussle not just with the steps, but with my new-found partner as well, as to who was going to be top dog. I was perfectly happy to be taught how to dance. The difficulty lay in the fact that I'm allergic to being bossed around, and Karen wanted to do a lot of bossing. To be fair, given the task ahead of her, it was quite justified. So there was a lot of banter. When we were asked by the camera crew who was in charge, we both chimed, 'I am!' at the same time.

The task in week one was relatively simple. Don't fall over, and don't come last. By the time we got to the big day, I knew the dance inside out. Karen had made the choreography deliberately simple – to accommodate my lack of expertise – although we did have quite a fun thing going, swapping a hat between us several times.

I don't care who you are, nothing can prepare you for the experience of going out and dancing on live television before an audience of ten million people. The pressure is unimaginable. You have to stand there in front of a red curtain, confronted by a camera and a producer dressed in black who is waiting to direct you onto the stage. The announcer issues his invitation: 'Will Bill Turnbull and his partner Karen Hardy please take to the floor for the Cha-cha-cha.' Well actually, no. Bill Turnbull will not. Bill Turnbull wishes to flee, with every fibre of his body, in the opposite direction. Bill Turnbull wants to go home. Bill Turnbull wants his mum.

But there's no stopping now. A last rictus grin for the camera. The producer whisks her finger stageward, and you're off. Round the corner, down the steps, take position. The band plays. This is it. Step-step-shuffle-shuffle-stop-back-rock. Two hundred moves to remember in the right order. It's a lot, but it's drilled in. The audience is clapping in time to the music. That's a good sign, they're enjoying it. I'm almost enjoying it myself as the feet go in all the right places and there are no mistakes. I end up on my knees (deliberately) before Karen prods me in the back with her toe and I fall forward, looking up at the camera. As the music stops, she is standing dominant with her foot on my back. Typical.

Big applause. Decent comments from the judges. Even before the votes start coming in, I know we're through to the next round. I feel like a million pounds. And Karen, performing in public for the first time in five years, is crying. Buckets.

We had twenty-four hours in which to bask in the comfort of having got through, before the next stage began. After a while, the week could be sectioned into different moods:

Sunday: Rest and relaxation.

Monday: Learning the first steps of a new dance. Never easy and requires a lot of concentration. But hey, we've got all week.

Tuesday: Still learning, and wondering if I'll ever get the hang of it. Still plenty of time though.

Wednesday: Most of the sections have been knocked in. It's coming together. But the weekend is beginning to loom ahead in the distance.

Thursday: The last full day of practice. Time to polish. Time to worry.

Friday: We take the dance to the studio at Television Centre. Waiting our turn to practise before the technical crew, we get to see what everyone else is doing. Sometimes that's a good thing; sometimes not. Beginning to feel mildly sick.

Saturday: Oh, crap. It's here again. A full day in studio. The whole programme is walked through with dancers and presenters. We get to perform with the band for the first time; sometimes that's a tricky moment as their production can be rather different from the recording we've been practising with. Everyone is friendly and cheerful. I don't know how they're feeling, but deep inside I'm beginning to churn. In my dressing room I find myself playing sad songs on the iPod. Probably not the best way to prepare, psychologically. And I feel tired, really sleepy. Karen says it's the nerves. Oh dear. Oh very dear. How did I manage to get myself mixed up in all this?

Part of the terror was induced by the judges. They would sit on their little platform overlooking the floor, dressed largely in black, as if already preparing to condemn the poor contestant pleading their case by performing a dance: Craig Revel Horwood, Arlene Phillips, Len Goodman and Bruno Tonioli. Since I took part, I've discovered that they are all honest, upstanding and even warm individuals. But when they're casting judgement on a routine you've given your heart and soul to for a week, they can come across as ogres. Well, to me at any rate.

The great comforters were the dressers. Yes, we had them. They were there to make sure that we got the right clothes on at the right time, and they were absolutely lovely. Putting your costume on for *Strictly* is rather more complicated than throwing on a T-shirt and jeans. Heaven knows what the girls had to do, but we boys had hurdles of our own to get over.

You'd think a costume for the Latin dances would be simple enough, wouldn't you? Just a shirt and trousers. But no. To put your shirt on, you have to step into your underpants, which consisted of a pair of lycra shorts, the kind that cyclists wear. This is because the pants are sewn onto the shirt, to make sure that in the full flight of the dance, the shirt doesn't come untucked to reveal expanses of (in my case) not terribly attractive-looking midriff. It also means that there's no awkward panty-line for the cameras to pick up on. Yep, my twenty-eight years of journalism had come to this. Panty-lines and sequins.

The ballroom outfit was rather more intricate, a kit of parts that had to be assembled with patience. Once the shirt was on, you had to proceed from the top, with the starched collar – bow tie already sewn on – attached to a starched white bib, which was in turn attached to a white cummerbund affair which acted as linchpin for the trousers. But these were no ordinary trousers. For ease of movement, the waist was very high – almost up to the chest, and hence the fly seemed to be about two foot deep. Dressing and undressing took much longer than usual, as did going to the loo. So if you wanted to pay a last-minute visit to calm the nerves, permission was required to make sure you had enough time.

It was indeed a strange new world, far away from what I was used to. I realised after a while that I had entered a new place called 'luvviedom', a rather nice territory where everyone you meet automatically treats you like a friend. You qualify for citizenship when you find yourself kissing women even though you are meeting for the first time. It's not unpleasant.

There was a huge amount to get used to in a short space of

time; not least the intimacy of the dance. Karen required me to put my hands and feet in all sorts of places they weren't used to going. First off, we had to stand much closer to each other than one does in normal society. Whenever we were standing together, or being interviewed, she would take my hand. All the dancers do the same with their partners. It's a very touchy-feely world. No wonder they're smiling all the time.

The first really big hurdle was practising the Quickstep, a dance I never liked as the steps don't match the rhythm and there's an awful lot of trotting and skipping required. At one point, Karen instructed me to take a step forward, between her legs. I moved forward half a pace. 'No,' she ordered. 'Really forward. Step between my legs.'

Oh. Dear. I was beginning to feel like the hapless, helpless hero in a French farce. This was going to require thigh to thigh contact. Oh dear, oh dear. Reluctantly I stretch my foot further forward.

'Now, William . . .' I wasn't sure what the next order was going to be, but I reckoned I wasn't going to like it much.

'THRUST!'

I shied away like a novice hurdler might if faced with Becher's Brook. But gradually, it became so routine as not to matter any more. After all, it wasn't as if I was being commanded to touch her up. I'd have to wait for the Rumba for that.

The Rumba. The dance of love. Yeah, right. I can't remember the exact story that Karen constructed for that one.

Something to do with love lost, found, lost, found and ultimately lost again, for there was a great deal of flitting to and fro, while I seemed to stand like a wet cow in a field and looking forlorn in the middle. Anyway, the plotline, Karen said, required me to place my hand on her bottom. This, presumably, during the ten seconds of the dance when we were entwined as lovers.

To be honest, I don't usually go around touching women's bottoms. At least, not since the court order took effect. So there was a natural reluctance on my part.

'Go on,' the instruction came. 'Grab my butt!' Oh well, here goes nothing. It was helped by the fact that Mrs T turned up for the next practice session, so I could show the move to her and see if she minded.

'Look, darling, I'm just putting my hand here. Is that all right?' Fortunately she understood it was 'necessary for the part' as the actors say just before they strip off all their clothes. And I must have 'just put my hand there' probably a hundred times. It was a tough job. But someone had to do it.

Clearly, it worked. The judges gave us a total of seventeen points out of a possible forty, and said it looked like Godzilla dancing with a praying mantis. A tad unfair. I thought we looked more like a bear trying to swat a bee.

What was keeping me in the competition, as we survived from week to week, wasn't talent by any means, but a touch of very odd luck. Something that could have knocked me out of the contest for good kept me in it instead.

There we were in week two, practising the Quickstep. It was all going fairly well, as we were putting the gloss on our moves in a gym in Slough. It's all glamour, this dancing business. Karen was unhappy with my *jeté*, a step where the front foot springs forward and the back foot is straightened. It was supposed to have the athletic grace of a gazelle; mine had all the spring of a carthorse.

So she made me improve my technique by leaping over some little platforms on the floor, the type they use for step exercises. All was progressing nicely until I landed badly on my left foot and went over on my ankle. It hurt, a lot. We tried moving on it a bit more just to see if we could dance it off, but it was no good. It was Thursday afternoon. We were in trouble. Fortunately, the video producers, on hand to record every scintillating second of our practice for the programme, had captured the moment of drama. As great moments of disaster go, it was hardly up there with the crash of the *Hindenburg*, but it did the trick.

If there's one thing the producers of *Strictly* do well – and they do an awful lot well – it's duty of care. The next morning I was transported up to Harley Street, to see a specialist who strapped it up tight, and gave me the green light to dance on the Saturday night. Amazingly, it worked. In fact, the whole episode worked very well in my favour. For as was repeatedly pointed out in the programme, it was a *horrific injury*, and I was being *incredibly brave*. And I suspect the resulting viewer sympathy kept me in the competition for a good couple of weeks longer than I would have been otherwise.

During this time, autumn was creeping on, and the bees were left to their own devices. I took Karen down to the farm for her first encounter, and enjoyed the idea that she was the one out of her comfort zone, for once. Especially when she shrieked: 'Bill! There are bees all over me!' They even featured on *Strictly*, when Chris Beale and a gang of loyal supporters from the Pinner & Ruislip Beekeepers Association gathered around a hive to wish me luck.

Indeed the bees – or at least their philosophy – were to help me one week, when we danced the Tango. I thought we'd performed a pretty neat little number to the tune of Doris Day's 'Perhaps'. It was one of the rare occasions when I felt we'd absolutely nailed it. But I had reckoned without the judges.

Craig was first: All very camp and passive, the dance had reminded him of Julian Clary. Then Arlene: No real passion, and she couldn't see our feet because the hem of Karen's flamenco skirt was so low. My inner bee had begun to buzz with irritation at Craig's 'camp' criticism. I was getting hammered in front of ten million people, and I wasn't going to take it lying down. So now the sting came out, and I let Arlene have it.

'No passion? Did you see the beginning of the dance? Did you SEE it?'

Oops. It's never a good idea to have a major strop in public, let alone on television in front of a huge Saturday night audience. I like to think it was due to all the painkillers I was having to take, because looking back at the footage later, I realised the judges had been right. The routine that I had thought was so amazing was really rather pitiful, almost embarrassing on my part. Fortunately the voting viewers had a good sense of humour, and kept us in the competition.

Our challenge the next week was the Paso Doble – the matador's dance. This allowed me to vent my frustration with the judges by stamping my feet all over the floor for hours on end. Cathartic possibly, but not great therapy for the ligaments that were trying to mend in my left foot. So while my head was still raging, things were beginning to heat up at the other end of the body. In order to keep things tight and immobile, the strapping had to stick adhesively to the skin. And every time the dressing came off, the skin became slightly inflamed, until one day it just gave up completely and came out in the most almighty red rash.

By now we were through to the next round, having danced what Arlene termed a 'Paso from Pinner'. Bearing in mind my membership of the Pinner & Ruislip Bee-keepers, I took formal exception to this, even though I'd been wearing an outfit that made me look like a bespangled chimneysweep. But on we marched.

Now it got interesting. We were trying to practise the Viennese waltz, complete with forty-seven dizzying spins, but the injury was really playing up. It was beginning to look as if I'd dipped my foot and lower leg into some scarlet paint, and felt like I was stepping around in a bucket of red ants. For the first time I did begin to worry about the injury, as opposed to the usual Thursday neurosis about the performance.

Instead of the regular trip to Harley Street, we had to go to Hammersmith Hospital for some emergency treatment. The doctor was a lovely Greek lady who had never heard of *Strictly Come Dancing* and had no idea what all the fuss was about. She administered a new dressing and a prescription for even more pills and then uttered the killer phrase:

'And no dancing. For several days.'

I summoned up my best thespian pleading. 'But you don't understand, doctor. I simply *have* to dance. People around the country are expecting it.'

'Okay then. But at your own risk.'

Phew! The nation could breathe again.

Quite understandably, the *Strictly* producers milked it for all it was worth. Would one of the celebrities be able to dance with a fearful injury? They even put us on air for a pre-dance interview, in which the hostess Tess Daly asked me if I was able to go out and dance. As I was all togged up in white tie and tails and had my shoes on, the answer was fairly obvious. It would have been a laugh if I'd said no. But instead I pronounced solemnly, 'We're here. We're here to dance.'

Afterwards Karen very generously suggested that it was the requirement to do the last-minute extra interview that broke my concentration and led to the terpsichorean tragedy that ensued. Maybe. Perhaps it was all the pills I was taking: anti-inflammatories, antibiotics and the rest. But I reckon it was just another case of Bad Ballroom, and a failure to concentrate.

It was all going so well (a phrase I got used to saying every week), until we entered the *fleckerl* right in the middle of the dance floor. This was a spinning manoeuvre which required lots of little tippy-toe steps while twirling one's partner with the tip of one hand. It was, to be honest, a devil to learn. But I'd practised it to death, and on the night had it down. As we came out of the last spin I thought, 'Hurray. That's done. The worst is over.' Silly me, of course it wasn't. The worst was yet to come.

As we came out of the spin, Karen looked at me with horror on her face and called above the music, 'We're on the wrong side of the floor! Follow me!'

I'd taken one turn too many, and we were now completely out of place and, given the intricate choreography, rather out of time. There was not a second in which to think as we charged around the floor, until halfway down the back straight it all fell apart. We broke contact, which for a waltz is rather like the wheels coming off a car. For a few bars we swayed helplessly as I tried to work out what I should do next. 'It's all over. We're going down. This is going to be our last dance, so leave her with something to remember.' I stepped forward and gave Karen a kiss on the cheek. What the hell.

The music – 'Golden Brown' by The Stranglers – came to an end. And I reckoned our chances of staying in the competition had been pretty effectively strangled by that performance. So you could have knocked me over with a feather when the judges awarded us twenty-seven – our second highest score so far. The drama of the foot had helped us through to the next round.

The next day – Sunday – it was really uncomfortable. I was at my wits' end as to what to do. How could I get back on my feet – literally – to learn the next routine? I turned to the healing power of honey; maybe the bees could help? Someone had told me that Manuka honey in particular could help soothe my burning sores, which felt, after the exertions of round five, like they were ready to burst. Manuka honey comes from New Zealand bees who feed on the flowers of the Manuka bush, to produce a honey that has extra anti-bacterial properties. It is said to be a useful treatment for everything from peptic ulcers and irritable bowel syndrome, to bed sores and chemical burns. It would surely then be more than a match for the rash on my left foot.

A trip to the supermarket later, I sat on the sofa and gently smeared lashings of this rather expensive product on the affected area, before waiting for the gentle balm of its healing powers to take effect. Today, I can't remember what it felt like when it did start working, perhaps because my memory is overwhelmed by the image of it. I don't know if it was an allergic reaction or just nature taking its course, but before my eyes the red sores gradually transformed into hard brown cells. The skin was bloating to form dozens of hard bubbles. In short, it looked as if I was turning into an alligator. This wasn't exactly what I had been expecting from the magic of Manuka, but there was no turning back now.

The creeping croc-mutation meant that I had to miss practice and make another trip to another hospital, and now I was getting depressed. Slowly but surely the knobbles of the alligator skin combined to form a vast blister right across my ankle and beyond. My left foot had taken on a life of its own, and all I could do was wait for it to be done with me. Fortunately our next dance was the aforementioned Rumba, a gentle affair. It didn't require any rapid movement, just a few steps, a lot of swishing and swaying, and me trying to look heartbroken. The rumba was the one dance I really wished I could do again after we'd performed it, largely because I'd forgotten a whole swathe of steps in the middle, and I wanted to show that I could actually complete a routine without the absent-minded swishing and swaying. Perhaps the excitement of the butt-grabbing had destroyed my concentration. With our pathetic score of seventeen we were, as the phrase went, 'bottom of the leader board'. Once again we were saved by the loyalty of the TV audience, but the writing was on the wall.

By the time Karen and I came to the waltz, what had felt like the world's largest blister was healing. The foot was no longer a problem. It was just my whole body letting me down now. Training eighteen hours a week or more for two months had taken its toll. Put simply, I was knackered, waking up as tired in the morning as I had been the night before. By now I was the oldest surviving contestant by a margin of at least ten years. The others – James Martin, Patsy Palmer, Colin Jackson, Zoe Ball and Darren Gough – were all younger and fitter, and still up for it in the seventh round. At training sessions, all I wanted to do was lie down. No amount of honey on my toast at breakfast could provide enough of a boost. Like a foraging bee at the end of summer, I was running on empty. And on the day of the dance, I knew it was time for me to go; Karen knew it too. Out on the floor, my head wanted to perform a waltz. The body, fuelled by adrenalin, was trying to do a Quickstep. The judges were kind. Even 'Craig the Cruel' said it was my best dance, before adding 'and probably your last'. So it was no surprise when Brucie read our names out as the couple who had to leave. It was, for the poor old body at least, a relief.

Looking back, I realised it had been the most exciting and fun-filled time of my life; an all-consuming rollercoaster of soul-ripping nerves, physical exertion, almost constant motion and more laughs than anyone deserves. And when it was over, it almost felt like being cast into outer darkness; like being a guest at the best party in the world and turning round to find the door slammed shut, with no way of getting back in. All of a sudden there was no more dancing. The practice sessions planned for the following week were cancelled. Those lovely endorphins, the natural high from taking so much exercise, stopped flowing. It felt like coming off a drug: the ballroom equivalent of 'cold turkey'. And it took a lot of getting used to.

I learned so much though; about how much you can achieve if you really set your mind to it. About the joy of the dance. The nature of celebrity. And how your bees really don't mind if you leave them alone. In fact, they probably prefer it.

Chapter Fourteen

FELLOWSHIP

*Where, at our journey's end, we discover what we
share with alcoholics and freemasons; and reveal
the spirit of the beehive.*

Many years ago, when I was a much, much younger man, I
took a holiday in Bermuda. I was living in New York at the
time, trying to make a living as a freelance journalist. It was
the middle of winter, when Manhattan is icy cold and no
matter how many layers you wear, the wind bites right
through you. I'd been seduced by the TV adverts promising
me warmth and sunshine and friendly local smiles. I was, so
to speak, 'between girlfriends' at the time, and needed to be
somewhere else to clear my head for a few days. Bermuda
wasn't far away and even on my feeble income I could still
afford a short break.

The odd thing about Bermuda in those days was that they would not let you in unless you had already booked somewhere to stay. You could fly all the way out there into the middle of the Atlantic, but you couldn't gain entry without a formal reservation. No sleeping on the beach here, thank you very much.

In my case, it was the usual answer. Of course I hadn't.

So what happened next was that the authorities would help you find a bed and breakfast somewhere. It didn't take long, and I was soon away in a taxi to my island paradise. Or rather, a single bed with a shared bathroom.

I am getting to the point here, honest. Of course, when you turn up alone, out of season, in a small B&B on a strange island and a modest budget, it can be a rather solitary experience. I had no friends and no contacts. In fact, after a day I wondered what the hell I was doing there.

But there was one other guest at the B&B, and he had no such problems. We compared notes at breakfast every morning. He too had flown in from the United States, and he didn't know anyone either. But he was busy every day going to meet new people, who were only too happy to welcome him into their homes and show him around.

The secret of his success? Alcoholism. He was a recovering alcoholic, and hence a member of Alcoholics Anonymous. It's possibly the most hospitable club in the world, and I mean that in the nicest possible way. For if, as an alcoholic, you bottom out and enter the twelve-step programme of AA, you are guaranteed the support of fellow members anywhere and everywhere. That's what makes the system work.

So all my fellow guest in Bermuda had to do was turn up, look up AA in the phone book, and bingo! His social life was sorted for the next week, while I hung around in bars making dribs and drabs of conversation with people, getting nowhere and feeling mildly miserable. It didn't seem right, somehow. I was almost tempted to start bingeing wildly for a day or two and then ask to join him and his mates, but it felt like too much hard work.

What struck me about this – patience now, I really am getting to the point here – was the fellowship that my companion had enjoyed. He could go anywhere in the world and make friends. It's something that not many people are privileged to experience. Freemasons can do it. Police officers. Members of fundamentalist sects. You have to be bound by a shared, sometimes religious fervour. Like a passion for bee-keeping. See, I got there in the end. Meet another beekeeper anywhere and you are guaranteed a conversation. Where do you keep your bees? How many hives have you got? Did you get any honey this year? What about varroa? And from there the talk may wander down dozens of different paths, as you discuss your various techniques and treatments, for in any beekeeping situation that are always different ways of doing things. For any question, there is rarely one right answer, and there are always plenty of wrong ones. You know right from the start that you and your companion are on the same side, walking in the same direction and generally with the same purpose.

The closest I've come to the feeling that my anonymous alcoholic friend in Bermuda must have enjoyed is in Northern Ireland, a place I got to know quite well in my reporting days. I used to get sent over as a backup for the regular correspondent, Denis Murray, either when he was away or if things got really nasty. If ever they were expecting repercussions from a particularly murderous bomb or shooting, I would be despatched on the next plane to help out. What usually happened then, when I turned up, was very little. The expected retaliation would not take place. Life would return to what, there, passed for normal, and I would be home again in a couple of days. As a result, in the newsroom they used to call me the Prince of Peace.

For a journalist, Northern Ireland was a great place to work, but it was immensely challenging. In those days correspondents reporting from abroad could have the freedom to assess a situation and write about it without worrying too much about what the locals thought of their reports. Northern Ireland, of course, was not abroad. And you could be sure that the parties involved – political and paramilitary – would be paying close attention to your every word. Sometimes the sensitivities were so delicate that reporting there felt like walking on eggshells in a minefield. If the eggs started to crack, there was no other place to go.

But the people were grand. I always thought what a great

place it could be to visit without the trouble. And eventually, it was.

I hadn't been to Northern Ireland for twelve years, since before the first IRA ceasefire, when I got the letter from the Mid-Antrim Beekeepers' Association. They had built a new bee hut in their apiary at Cullybackey, and would I come and open it for them? We've all heard of celebrities so hungry for publicity that they'd turn out for the opening of a paper bag, but I'll bet they never attended the opening of a hut. This would be a first. So it didn't take long for me to accept the chance to see Northern Ireland again, only this time at peace. And the idea of opening anything in a place called Cullybackey was attraction enough in itself.

They spoilt us rotten. I now know what it's like to be a royal, for that's how we were treated on our arrival. The reception committee greeted us warmly and then led us past a long row of respectful spectators who all applauded as we went by. I almost felt I should be walking with one hand behind my back, stopping occasionally to say hello to someone with a regal, 'Have you been waiting long?' There was a ribbon and some scissors and a photographer and a speech, and I can't remember a word I said.

All that was missing was some bees. There was not a single one to be seen, let alone a hive. This was because they'd recently had a foulbrood alert in the area, and hadn't been able to bring any bees on site as a result. So we had a large crowd, a grand ceremony, and a great big hut. But the stars of the show were missing. All we could do was look out of the large windows built especially for viewing the bees at work, and take in the flowers. Still, we had a grand party, with a roast pig and even lobster thermidor for the guests of honour.

One thing leads to another and it wasn't long before I received another invitation from Northern Ireland, this time to become president of the Institute of Beekeepers, set up to promote education and better beekeeping methods. Quite how I should qualify to represent any organisation on either of those counts was beyond me. Had they known the full extent of my apicultural misdemeanours, they might have reconsidered their invitation. But as I had never been president of anything before – and how many of us have? – I jumped at the opportunity. My duties required me to travel to a different part of Northern Ireland once a year for the INIB annual conference, bang a gavel at the beginning and the end of the day, wear a heavy chain of office round my neck throughout the proceedings, and deliver a speech worthy of my office to get the ball rolling. It's the last task that has always been the most troubling. What could I possibly say to an audience of two hundred beekeepers, almost all of whom would probably be more experienced and definitely more proficient at the noble art than myself?

So in my first opening address as president, I decided to come clean. The beekeepers of Northern Ireland were good folk who deserved, for better or worse, to know what they had let themselves in for. I let them have it, the catalogue of my crimes: carelessness, stupidity, wanton foolishness, murder and general beeslaughter. It was warmly received. Perhaps because they were just being polite, or perhaps because they just couldn't believe that anyone could be quite that bad, and that I must have been kidding. Only an idiot or a wise man would be foolish or brave enough to recount all of his faults in front of an audience, so they erred on the generous side.

What I love about the Institute is that it's a great unifier, where beekeepers from all over the six counties come together to learn more about their craft. Not so long ago, it wasn't easy to do that. In fact it just didn't happen. But today there are no politics, and no policies to agree on. It's just about the bees.

The point of the INIB is to advance the cause of bee-keeping through education. And these days it seems the cause needs all the advancement it can get, in the face of its adversaries: varroa, tracheal mites, the small hive beetle, Israeli acute paralysis virus, European foulbrood, Kashmir bee virus, deformed wing virus, dysentery, and pesticides. 'Bee health is at risk and, frankly, if nothing is done about it, the fact is the honeybee population could be wiped out in ten years.' That's a pretty grim warning, the sort of apocalyptic vision you might expect from a keen environmental activist. In fact, they are the words of Lord Rooker, the government's environment and rural affairs minister, in 2008.

There's a certain amount that can be done through research; getting to the root causes of Colony Collapse Disorder and the other ailments. Environmentally, perhaps we're all beekeepers now. Anyone who has a garden, or even just a window box, can do something to help by planting the sort of flowers that honeybees and bumbles can feed on: geraniums, foxgloves, hebe and lavender to name a few.

But that doesn't mean we should all be keeping bees. It's a complicated business, and while it can be immensely rewarding, it's not particularly easy. Too many inexperienced novices trying to keep too many hives would be bad for everyone. There's only room for a certain number of Bad Beekeepers in the world, so form an orderly queue.

But here's something to think about while you're waiting. How much do you like being stung? Or rather, how much do you like the risk of being stung? At one conference I attended, beekeepers were asked if they would like to have bees that simply didn't sting any more. Only a couple of people in the audience raised their hand.

There's a reason for this; it's what separates beekeepers from everybody else. It may seem weird, but most of us don't mind being stung. We appreciate that it's a clause in the contract. The pain is part of the package, the price you pay. As I said before, in a way it's where the fun is. When you become – and stay – a beekeeper, you join the fellowship of people who share that characteristic. No wonder people think we're odd.

Ultimately, though, it's all about the union with the bees. Forget about the honey, and the candlemaking, and the mead and all the other little bits and pieces you can harvest from the hive, like pollen and propolis. The real joy for me is just in keeping the bees, literally. From the moment you lift the lid off the hive and peer inside, you are entering another world, and leaving your own. When the sun is shining, you will find the residents in their thousands sitting on top of the frames, calmly going about their business. They may barely notice that you are there. And if they do, if they are placid, they will let you share the moment. Once you delve inside to check that they are healthy, for a while your world with all its human concerns will drift away. The phone will not ring. Even if you were foolish enough still to be carrying your mobile, sealed inside your suit and veil, you cannot answer it anyway. You are totally absorbed by what is happening in the city, the 'apisopolis', before you. You look to see how many citizens there are, and then how they are living. Do they have enough honey? Are there stores of pollen? And crucially, is the colony repopulating? Are there new bees being born? Are there eggs and larvae in abundance? And where is She, the most important one of all? Once the queen is found, you can enjoy a moment of relief and satisfaction that whatever other issues there may be, the future of the colony is assured, for the time being at least.

And then you look for other signs, for the things you hope not to find. Like the tiny red varroa mite riding on the back of some unfortunate worker, or bees with wings chewed in irritation, or some indication of disease that needs attention.

Once you've established what's going on, you work out what to do next. Do you need to make more room to stop them swarming? Should they be split into separate hives? Or joined together with another? Do they need a dose of syrup to keep them going? A dozen questions may be raised by a visit to a single colony, and for every issue there's at least one choice that is right and probably several that are wrong. Trouble is, you only find out if you're correct once you've seen the consequences some time later. But that, too, is where the fun is; in the never-ending process of discovery and learning. The only fact about beekeeping that you can be sure of is that you'll never know them all.

And when you've finished, you put the hive back together, and leave the bees to their secret life in the darkness, and come back up into the world of light. But for a while you have been away – even though you've been standing for some time on the same spot – away to a place that, in the grand scheme of things, not many get to visit. And it can feel like magic. Even for a Bad Beekeeper.

Sometimes you might just wonder if it's all worth it – the stings, the sweat, the blood and the tears, the trouble, the mess, and the frustration. When you've done everything you can to keep the little darlings alive through the winter, and they still die on you. When your favourite colony succumbs to some dreadful lurgy that you failed to detect; when nature conspires against your best efforts and rains all day, every day, through the peak of the summer; when you open up the hive to find that not only is there no honey, but there are no bees either; then you might ask yourself, what really is the point?

The answer, for me at least, lies in that sense of peace that you can only find with your head down a beehive. And in the hope that in trying to master the disciplines of beekeeping, I may have become vaguely acquainted with a virtue or two; though I think it'll be a while before we're bosom buddies.

But enough of this dreamy philosophising. It's late autumn and there are three boxes of honey on the kitchen floor which have been waiting to be processed for the past six weeks. The extractor is sitting outside in the rain, as if asking for permission to come back in. For members of the Bad Beekeepers Club, the work is never done. That's what makes us what we are.

And what in heaven's name is that dog licking now?

A Beekeeper's Blessing

May your colonies be healthy and plentiful
May your supers be overflowing
And may your swarms always be someone else's bees!

Acknowledgements

This book would not have been born but for the inspiration of Sylvia Tidy-Harris, who raised the idea on the ferry to Sark, and Jonathan Conway, who made me believe that I really could write a book. Turns out he was right.

I would never have been even half a Bad Beekeeper but for the patience and wisdom of Christopher Beale, who also kindly checked that I hadn't made a complete fool of myself with my facts.

Thanks also to all at Little, Brown: especially Antonia Hodgson and Hannah Boursnell.

And to the family, for putting up with the bees in the first place.